fascinating SHELLS

An Introduction to 121 of the World's Most Wonderful Mollusks

Andreia Salvador

The University of Chicago Press

To Roberto

The University of Chicago Press, Chicago 60637

Published 2022
Printed in China

31 30 29 28 27 26 25 24   2 3 4 5

ISBN-13: 978-0-226-81913-6 (cloth)
ISBN-13: 978-0-226-81981-5 (e-book)
DOI: https://doi.org/10.7208/chicago/9780226819815.001.0001

First published as *Interesting Shells* in 2022 by the Natural History Museum, Cromwell Road, London SW7 5BD.

Photography by Kevin Webb.

Library of Congress Cataloging-in-Publication Data
Names: Salvador, Andreia, author.
Title: Fascinating shells : an introduction to 121 of the world's most wonderful mollusks / Andreia Salvador.
Description: Chicago : The University of Chicago Press, 2022. | First published as *Interesting Shells* in 2022 by the Natural History Museum, London.
Identifiers: LCCN 2021025584 | ISBN 9780226819136 (cloth) | ISBN 9780226819815 (ebook)
Subjects: LCSH: Natural History Museum (London, England)—Catalogs. | Shells—Identification. | Shells—Pictorial works.
Classification: LCC QL404 .S34 2022 | DDC 591.47/7—dc23
LC record available at https://lccn.loc.gov/2021025584

Reproduction by Saxon Digital Services, Norfolk, UK
Printed by Toppan Leefung Printing Limited, China

The Museum

London's Natural History Museum is not only a tourist attraction, but also a world-class research institution that employs over 300 scientists and houses many of the world's most important taxonomic collections. The Museum's mollusca collection is one of the most comprehensive and significant in the world and has over 8 million specimens. These specimens are key to telling the history of collecting, the science of taxonomy and the human desire to understand the natural world.

The author

The author is the Senior Curator of Mollusca at the Natural History Museum, London. Her interests include history of collections, voyages and expeditions in malacology. She has been collecting seashells since she was a child, and her favourite one is the mini museum of curiosities, the carrier shell.

Acknowledgements

The author would like to thank: Kevin Webb, Roberto Portela Miguez, Jonathan Ablett, Tom White, Suzanne Williams, John Taylor, Chong Chen, Graham Oliver, Manuel Malaquias, Barna Páll-Gergely, Marta Ferreira, Susana Celestino, CIEMAR and NHM Publishing.

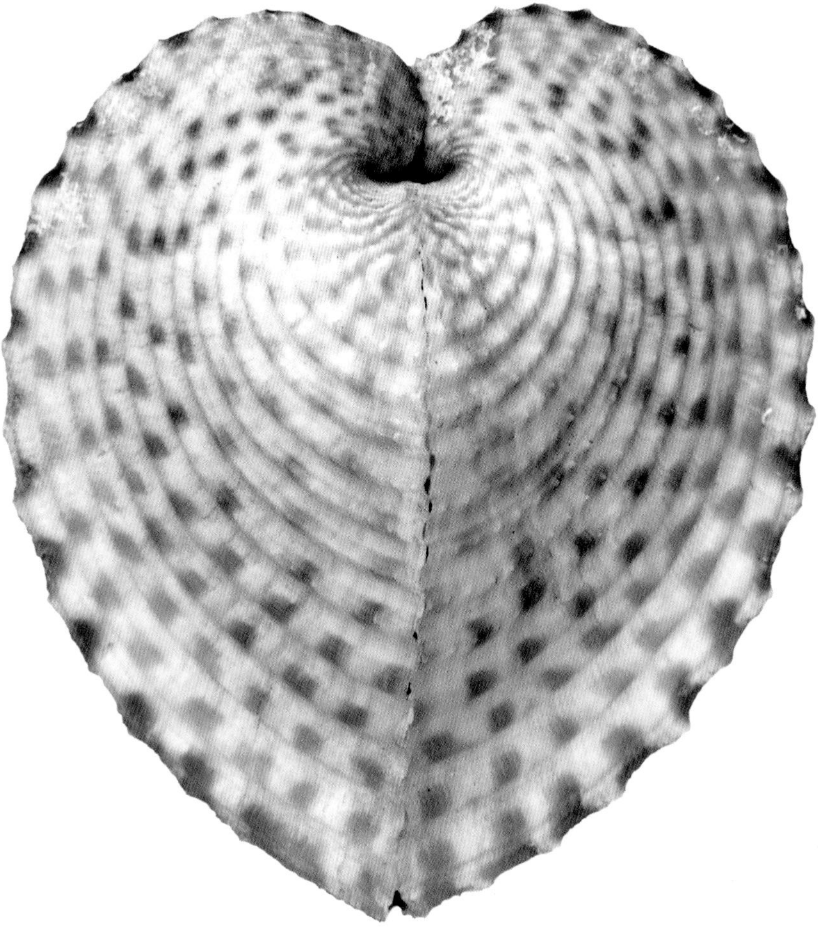

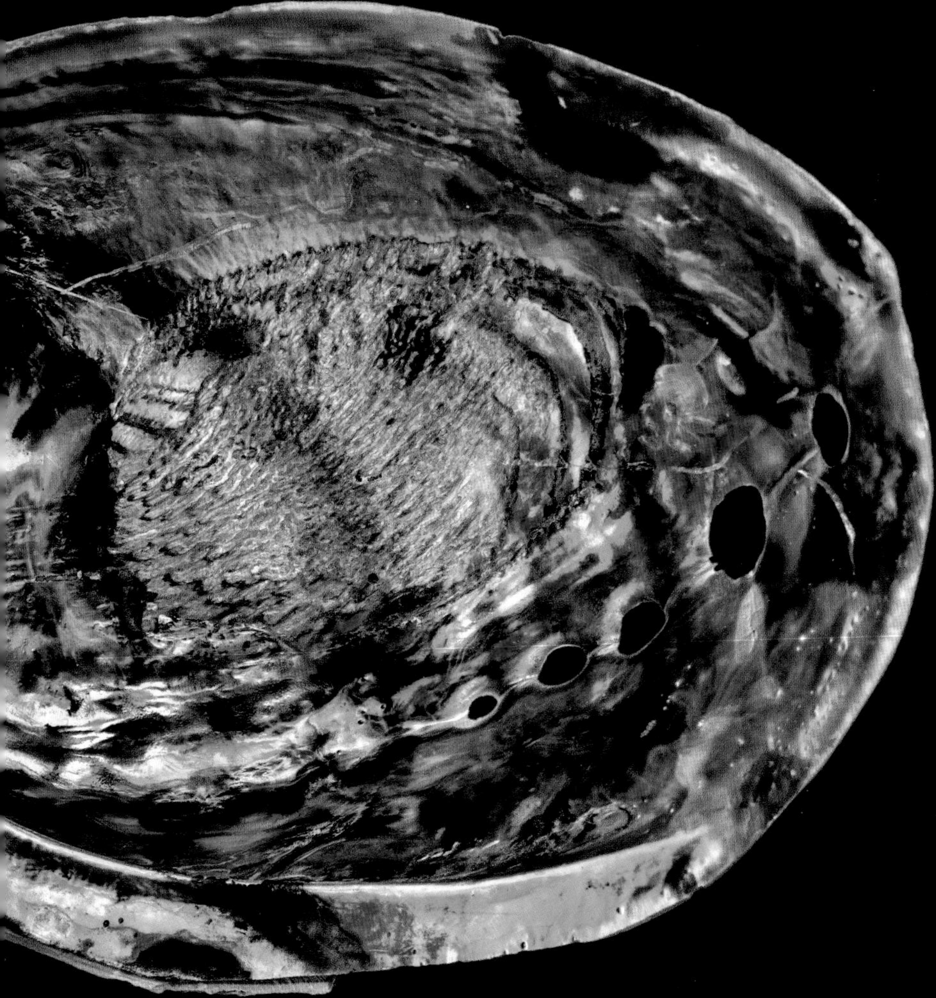

Hundred-eyed cowry

Arestorides argus

Each animal or plant is given a name that consists of two parts. The first part is the genus to which a species belongs, and the second part is the species name. This is known as the binomial naming system and was devised by Swedish scientist Carl Linnaeus (1707–1778). Linnaeus himself described this cowry in 1758. The numerous brown rings covering the shell of the animal reminded him of Argus Panoptes, a hundred-eyed giant in Greek mythology, whose name Panoptes meant 'the all-seeing one'.

Distribution
Indo-Pacific

Size
75 mm (3 in)

Common distorsio

Distorsio anus

This species is one of the most distinctive of all gastropods with its singular almost grotesque shell. It is characterized by its extremely distorted coiling with asymmetrical, strangely developed whorls on the spire. This appearance is because old varices remain in place as the animal grows, causing the whorls to have to grow over the expanded varix of the earlier whorl. A whorl is one complete turn of a spiral shell. A varix (plural varices) is the rib-like thickening, which is a former outer edge. A varix is formed at a rest period in the growth of the shell, and usually indicates an interruption in growth and a reinforcement of the shell edge. This distorsio occurs commonly in shallow water, inhabiting reef substrate under corals.

Distribution
Indo-Pacific

Size
70 mm (2 ¾ in)

Zigzag nerite

Vittina waigiensis

These colourful freshwater snail shells are a wonderful example of the diversity in molluscan shell colour and, although the pattern on each is different, they are all representatives of the same species. Such variable appearance within a species is known as polymorphism. The advantage of looking different is possibly very simple. If they all looked the same, predators would only have one pattern to recognize when hunting. With many different patterns, the zigzag nerite confuses predators by making it difficult to establish a consistent search image of the prey.

Distribution
Southwest Pacific

Size
15 mm (½ in)

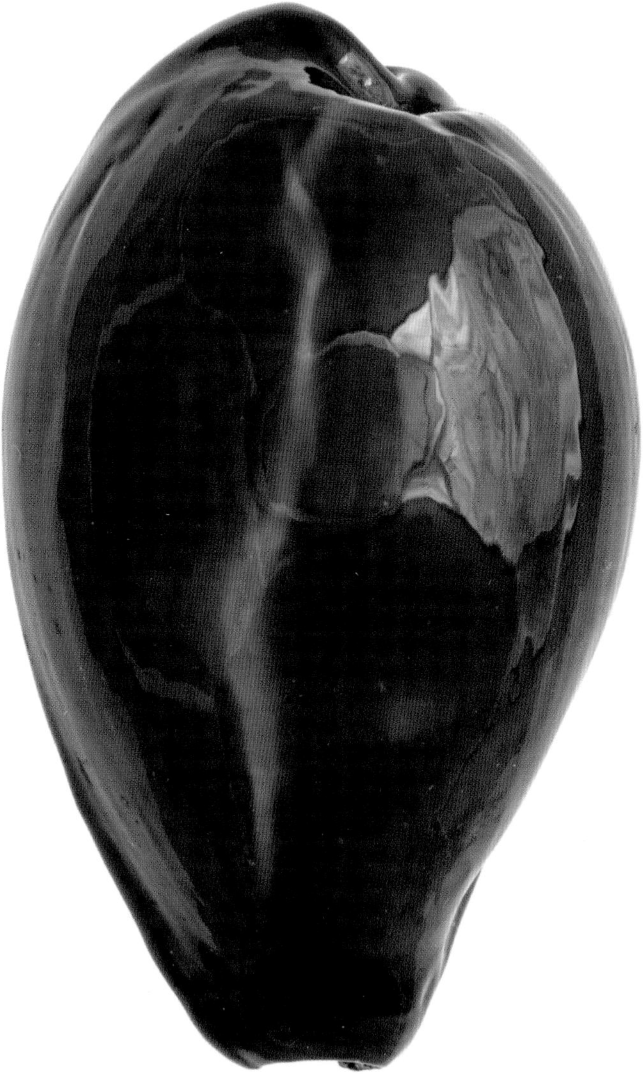

Tanned cowry

Erronea adusta

The polished surface of the cowry shell is produced by the action of the extensible bilobed mantle, an anatomic structure in molluscs that secretes calcium carbonate and conchiolin to create a shell. In cowries, when the mantle is extended, it covers the whole shell; a sinuous line often marks the position where the two lobes meet. The mantle, which is also responsible for the deposition of callus and pigments, protects the shell against parasites, and repairs cracks and holes caused by erosion or predators. In many species the mantle serves as camouflage, obscuring the shell's pattern completely and blending it in with its surroundings.

Distribution
Indian Ocean, Melanesia

Size
45 mm (1¾ in)

Beech cone

Conus betulinus

Cone shells are a favourite amongst collectors; they have a distinctive and easily recognizable shape but vary enormously in colour and colour pattern. This large group of marine gastropods is found worldwide in tropical and subtropical seas, with diversity reaching its highest in the tropical Indo-West Pacific region. They live in sandy regions or among rocks or coral reefs. Cone snails are predators and highly specialized in their feeding – they hunt and eat marine worms, fish or other molluscs. They immobilize their prey with a harpoon-like radula that injects a venom called conotoxin that quickly paralyzes their prey.

Distribution
Indo-Pacific

Size
60 mm (2 ½ in)

Spiky trophon

Enixotrophon carduelis

HMS *Challenger* was the first oceanographic research vessel and carried out the first major attempt to study the deep sea. She circumnavigated the globe from 1872 to 1876 with a crew of international scientists, who collected data from 362 oceanographic stations, investigating every physical and biological aspect of the oceans. The research resulted in 600 cases of material that took 76 specialists approximately 20 years to analyze, identify and describe. Consequently, more than 4,700 new species of marine life were discovered. This expedition demonstrated, once and for all, that animal life was to be found even in the very deepest part of the ocean. The spiky trophon was discovered and described by Reverend Robert Boog Watson (1823–1910), a Scottish malacologist responsible for reports on gastropods and scaphopods collected during the voyage.

Distribution
Australia, Tasmania, New Zealand and Indonesia

Size
30 mm (1 ¼ in)

Glassy nautilus

Carinaria cristata

This shell is extremely thin, fragile and translucent, as if made from the finest spun glass. It was thought to be the shell of a small nautilus, hence its popular name, but it is actually the shell of a pelagic snail, a heteropod that lives in the open sea. Heteropods are holoplanktonic, meaning that they spend their entire lives in the water column as part of the plankton. They live just below the surface to around 500 m (1,640 ft) deep. They are active predators, feeding on a variety of zooplankton, including other planktonic gastropods, small fish and crustaceans. They have complex eyes, which enable them to detect and pursue prey. The glassy nautilus is the largest heteropod of all.

Distribution
Indo-West Pacific

Size
70 mm (2 ¾ in)

Sun carrier shell

Stellaria solaris

The species name of this beautiful pale shell is translated as ‘of the stars, of the sun’. The animal lives on tropical sandy or muddy bottoms on the continental shelf from 18 to 200 m (60 to 655 ft). It feeds directly on the sediment, filtering out detritus and foraminifera. The remarkable symmetry of its shell is enhanced by its 10 to 20 long, narrow, blunt spines. These increase the effective area of the shell and improve its stability. By creating its own projections, this marine snail can use them like stilts to raise itself off the sea floor.

Distribution
Indo-Pacific, Red Sea, Persian Gulf

Size
105 mm (4 in) with spines

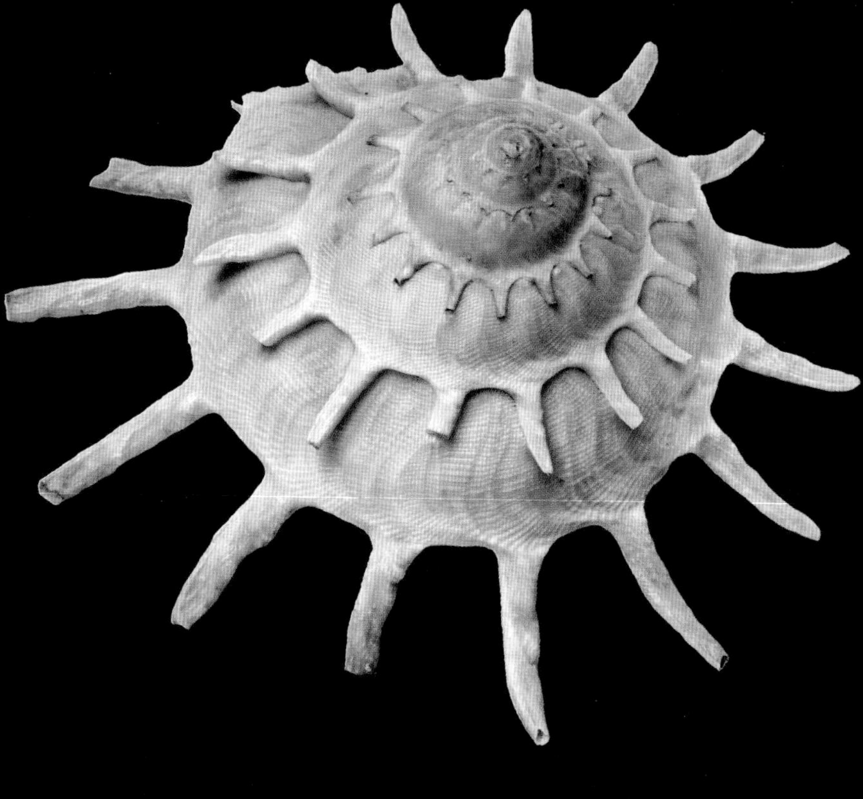

Violet sea snail

Janthina janthina

Violet sea snails live their entire adult life drifting upside-down with the currents on the surface of the open ocean. They make a raft by trapping bubbles of air in a jelly-like mass that is attached to the foot. The mucus used to produce the raft seems to be chemically stable so it is long-lived and the bubbles do not burst. The snail's violet shell is paper-thin, extremely light, and reverse countershaded because of its inverted position in the water. Countershading is a method of camouflage, where one side of an animal is dark and the other side is light. By having two different shades of purple, the shell blends with its surroundings, making it difficult for predators to see it from above or from below. These pelagic snails are carnivores and feed on hydrozoans, such as the Portuguese man-of-war. Since these snails cannot swim, they are entirely dependent on the wind and current to bring them into contact with prey and mates.

Distribution
Worldwide in the warm waters of tropical and temperate seas

Size
40 mm (1½ in)

Map cowry

Leporicypraea mappa

There are around 200 described species of cowries living today. Widely distributed in tropical and subtropical marine ecosystems, these gastropods live mainly in shallow water on coral reefs. Most hide in crevices and beneath boulders and slabs during the daytime, and are active only at night. Nearly all species are primarily carnivorous though some feed on a variety of organisms including algae. Cowries also have predators - some fish specialize in turning rocks and breaking coral to seek them out, and octopuses and other molluscs can drill holes into the shell to kill and eat them.

Distribution
Western Pacific

Size
80 mm (3 in)

Precious wentletrap

Epitonium scalare

The precious wentletrap was once considered to be a very rare shell. It is unique because the whorls are not tightly joined together as is the case with most spiral gastropods. Instead, the shell is held together by a series of white ridges, called varices, leaving gaps between the whorls. In the 17th and 18th centuries, at the height of their popularity, these shells would fetch several hundred pounds apiece and were so in demand that they were counterfeited using rice flour. However, it is said that the fraudulent shells were soon detected by the owners who dipped the fake shells into water to clean them, only to watch their precious wentletraps become worthless blobs. Though no longer considered rare, they are still greatly admired and appreciated by collectors, but their prices nowadays are a modest fraction of earlier times.

Distribution
Indo-Pacific

Size
65 mm (2 ½ in)

Lister's conch

Mirabilistrombus listeri

Martin Lister (1639–1712) was an English naturalist, physician and author of the first comprehensive study of conchology *Historiae Sive Synopsis Methodicae Conchyliorum* (History of Molluscs) (1685–1692). This significant work consists of over 1,000 copperplates portraying shells and molluscs collected from around the world. Susanna (1670–1738) and Anna (1671–1700) Lister, Martin's daughters, produced the illustrations and engravings of this book – the sisters were 15 and 14 years old respectively when the first part was published in 1685. The Lister family illustrated this species at the end of the 17th century, more than 150 years before it was described and named. In 1852 Thomas Gray described the species illustrated here as Lister's conch, after Martin Lister 'whose work is a lasting memorial of unwearied diligence and perseverance'.

Distribution
Northwest Indian Ocean, Andaman Sea

Size
100 mm (4 in)

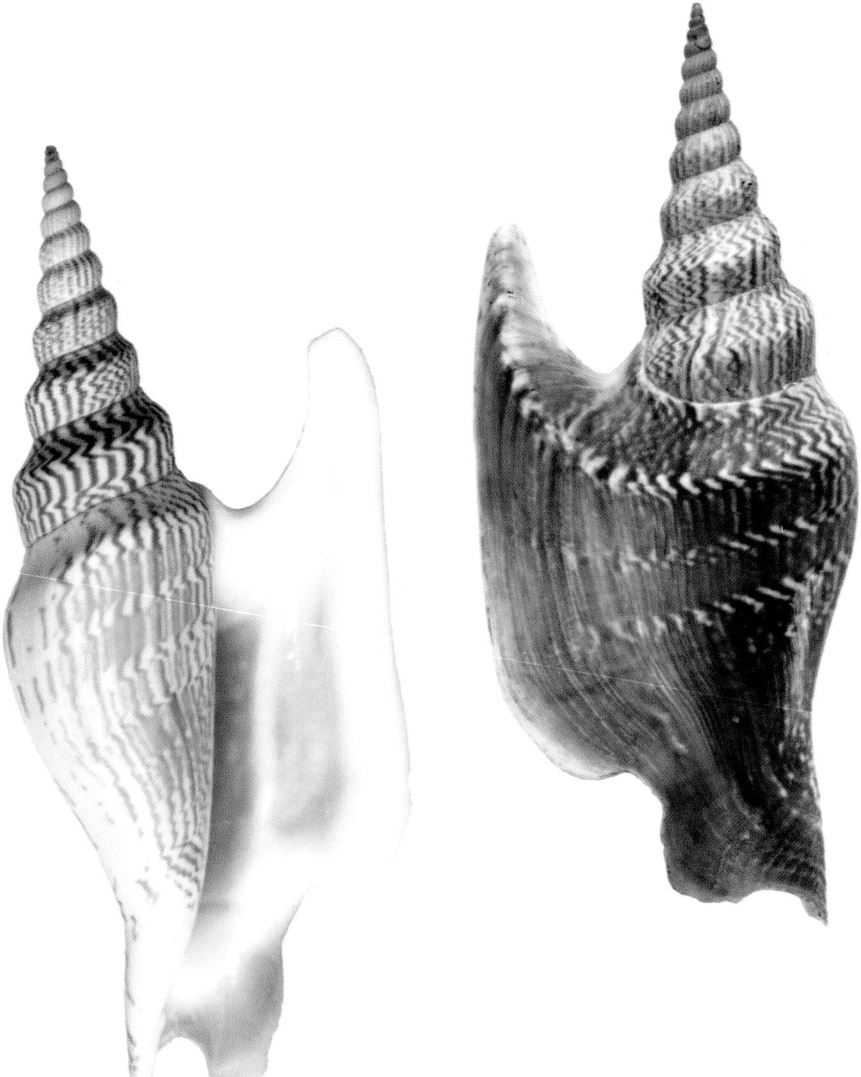

Amber snail

Powelliphanta hochstetteri

Snails have evolved to feed on a variety of food items - some are generalists whilst others are more specialized. Most land snails are herbivorous, eating leaves, roots, seeds, fruit, mushrooms, moss, algae and lichens. Others are carnivores and prefer to eat other snails, earthworms or rotting meat. Most gastropods have a rasping tongue-like organ called a radula, full of microscopic teeth that rub the food against the jaw like a cheese-grater. The amber snail, for example, eats earthworms and sucks them up through its mouth just like we eat spaghetti. Inside the mouth, the radula scrapes chunks of flesh into the stomach for digestion. They are among the largest land snails in the world, reaching up to 90 mm (3½ in) in diameter, and are also one of the most threatened of New Zealand's invertebrates.

Distribution
Endemic to the Marlborough and Nelson provinces of the South Island of New Zealand

Size
50 mm (2 in)

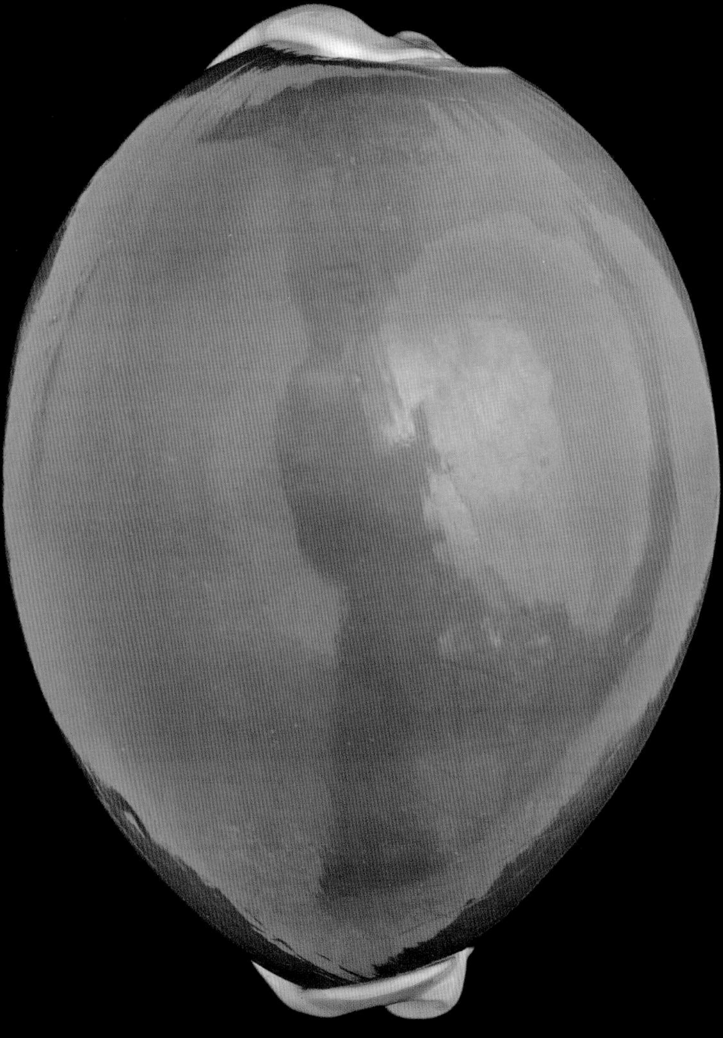

Golden cowry

Callistocypraea aurantium

The aesthetic appeal of cowry shells makes them a favourite among shell collectors. Their domed, highly glossy outer surface and often beautiful patterned shell make them easily recognized among marine gastropods. Since prehistoric times, cowry shells have fascinated humans, so unsurprisingly many have been recovered from settlements, graves and spiritual sites all over the world. The legendary golden cowry is one of the most famous and, although not considered as rare as it used to be, its shell is still highly prized. Its species name *aurantium* is of Persian origin, meaning orange fruit.

Distribution
Central and Western Pacific

Size
90 mm (3 ½ in)

Giant sundial

Architectonica maxima

Sundials have a characteristic and identifiable shell, with a very compact conical shape and a wide, deep umbilicus – the central cavity at the base of the shell, around which the whorls coil. They have worldwide distribution, mainly in subtropical and tropical waters, and occur at all depths, from the intertidal to the abyssal. They have a long-lived, free-swimming larval stage that can cover great distances, moving with the ocean currents. This accounts for the wide distribution of some species. The giant sundial is the largest member of the sundial family. It is a carnivore that feeds on sea anemones and coral polyps, and lives on sandy and muddy substrates at a depth of 10 to 165 m (33 to 540 ft).

Distribution
Indo-Pacific

Size
60 mm (2 ½ in)

Egg cowry

Ovula ovum

Beliefs in the mystical properties of shells are common across cultures. Since early times, the shells of marine snails have been used in some cultures as a symbol of status or representing good fortune - acting as charms against evil forces, against sickness or to favour fertility. The beautiful white shells of the egg cowry have been adopted by the people of Melanesia and Polynesia as a symbol of fertility and are used to decorate their huts, canoes and fishing nets. They are also used as ornaments and tribal symbols in bracelets and necklaces.

Distribution
Indo-Pacific, Red Sea

Size
110 mm (4 ¼ in)

Imperial volute

Cymbiola imperialis

Margaret Cavendish Bentinck, second Duchess of Portland (1715–1785), was a British aristocrat and collector of natural history and art. Her collection of shells and other natural objects was considered the finest in England and rivalled the best in Europe. The Portland collection was dispersed widely in 1786, after her death, in a renowned natural history auction that lasted 38 days. This species of volute was described for the first time in 1786 by Reverend John Lightfoot (1735–1788), as he catalogued the collection for sale. Only a few of the Duchess's shells can now be traced in museum collections.

Distribution
Philippines

Size
105 mm (4 in)

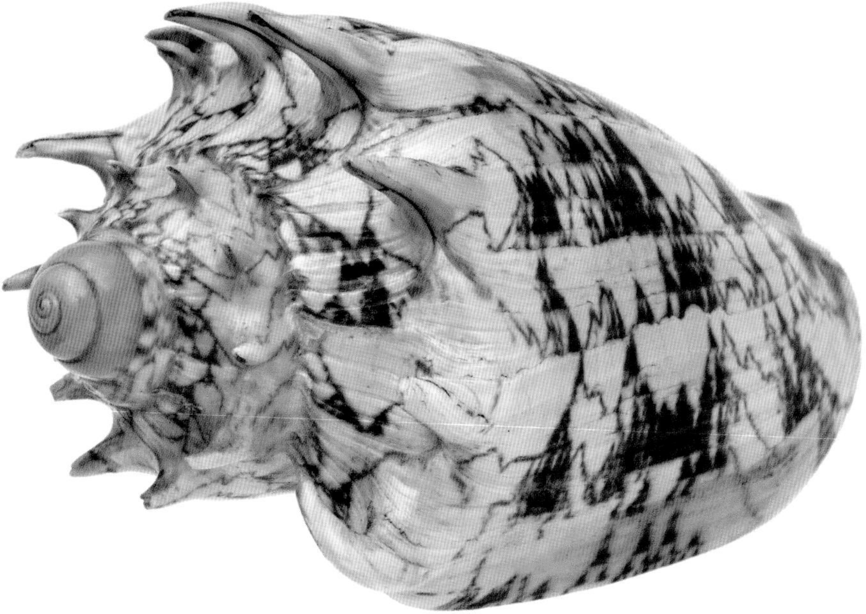

Purple dye murex

Bolinus brandaris

For thousands of years, people have known how to extract dye from molluscs. The Phoenicians of Tyre, in southern Lebanon, perfected the process for making a dark reddish-purple dye, known as Tyrian purple. This long-lasting dye was taken from several species of whelks including the purple dye murex, *Bolinus brandaris,* and the banded dye murex, *Hexaplex trunculus*, which are found on the Mediterranean and Atlantic coasts. When the mollusc is disturbed, it retreats into its shell and ejects a fluid, which initially is colourless but, once exposed to sunlight, gradually turns from yellow to green to blue and finally to a purplish red. The dye was expensive - for one gram of dye, more than 10,000 snails were needed. For this reason, the dye was reserved for the garments of individuals of the highest rank. Nowadays, synthetic versions of these pigments are manufactured.

Distribution
Portugal, Spain, Morocco, Mediterranean Sea

Size
70 mm (2 ¾ in)

Japanese wonder shell

Thatcheria mirabilis

This elegant shell, with its flat-topped whorls, is one of the world's most distinctive. Its shape is so unique that when the first specimen was discovered and described as a new species, many believed it to be just a distorted shell. More than half a century passed before any more specimens were found, proving evidence of the validity of the species beyond doubt. This species lives on fine sand and muddy bottoms at depths of 60–600 m (195–2,150 ft) and feeds on marine worms.

Distribution
Japan to Australia

Size
90 mm (3 ½ in)

Jewel slit shell

Perotrochus quoyanus

Slit shells are considered living fossils because they look like their extinct relatives and have changed little in millions of years. This group was thought to be extinct until the rare slit shell, *Perotrochus quoyanus,* was discovered in the French West Indies in 1856 by French naval officer, Commandant Beau. The fact that they only occur in deep water makes these animals difficult to collect. However, as the technology and techniques to explore deep habitats advanced, more slit snails were rapidly discovered around the world. They are named for the narrow slit in the middle of the last whorl, which serves as an opening for the passage of water and waste products.

Distribution
Caribbean Sea and Lesser Antilles

Size
46 mm (1 ¾ in)

Banded grove snail

Cepaea nemoralis

Since the arrival of Europeans, many species of molluscs have been intentionally or accidentally introduced into the United States. However, knowing the exact moment of introduction is unusual. William G. Binney (1833–1909), an American malacologist, collected hundreds of *Cepaea nemoralis* snails, one of the most common species of land snail in Europe, whilst in Sheffield, England. In 1857 he returned to the United States where he released them into his garden in Burlington, New Jersey. According to Binney, the experiment was a success, the snails thrived and their numbers increased rapidly. This is the earliest recorded introduction of *Cepaea nemoralis* into North America. The animal is currently found in Virginia, New York, Ontario and Massachusetts.

Distribution
Europe, North America

Size
20 mm (¾ in)

Carrier shell

Xenophora pallidula

This marine gastropod has a peculiar and charming behaviour – it collects and attaches foreign objects to the edge of its shell as it grows, including corals, other mollusc shells, pebbles, charcoal, fossils, and even man-made objects like broken glass, coins or plastic. It is unclear why the animal cements these bizarre objects to itself, although it may be for camouflage, increasing shell strength or protection. Whatever the reason, the result reflects a sample of their habitat, a mini collection that they carry with them all the time. It is the reason for its name, *Xenophora*, which is derived from two ancient Greek words meaning 'carrier of foreign objects'.

Distribution
Indo-Pacific

Size
70 mm (2 ¾ in) without the extra shells

Turban top shell

Gibbula magus

Top shells are marine snails named for their resemblance to a child's spinning top. They are a large group with worldwide distribution, found on rocky shores and on coral reefs, where they feed on seaweed. The inner part of their shell consists of a thick layer of mother-of-pearl. In older individuals, the outer layer is often eroded, leaving the mother-of-pearl exposed, especially at the tip. Some of the larger species of top shells are still collected for jewellery and for manufacturing items such as buttons, though this industry has been replaced by plastic and nowadays the demand is not as high as it was 100 years ago. The turban top shell is a common European species and most shells are cream-coloured with a pattern of pink or red lines. They can be found from the intertidal to depths of up to 70 m (230 ft) on rock, muddy gravel and muddy sand.

Distribution
Northeast Atlantic and Mediterranean

Size
30 mm (1 ¼ in)

Miraculous diplomat snail

Plectostoma mirabile

These miniscule land snails from Borneo, no bigger than a letter on this page, look like blown glass. Their fragile shells are almost invisible to the naked eye, but under the microscope each species has a unique shape, crafted in short spines, twists and turns with a distinct trumpet-shaped opening. The shape and number of spines vary, depending on where the snails live and, intriguingly, possibly even from what angle different predators attack. This tropical rainforest animal lives on limestone outcrops in northern Borneo, and they have an extremely limited range – entire populations can be limited to just a few rocks. Anthropogenic threats, such as quarrying, fire and logging, have destroyed much of the habitat of these unique land snails and, for this reason, they are now facing an extremely high risk of extinction in the wild.

Distribution
Borneo

Size
3 mm (0.1 in)

Donkey's ear abalone

Haliotis asinina

The shells of the abalones or ormers are characterized by having a single row of holes, called tremata. These holes are used during respiration, in which water is drawn over the animal's gills from underneath the shell and exhaled through the holes on the top of the shell. Sperm and eggs are also released into the water through these holes. As the animal grows and extends its shell, the holes continue to form - the oldest ones are gradually sealed up and new holes are made near the edge of the shell.

Distribution
Indo-West Pacific

Size
95 mm (3 ¾ in)

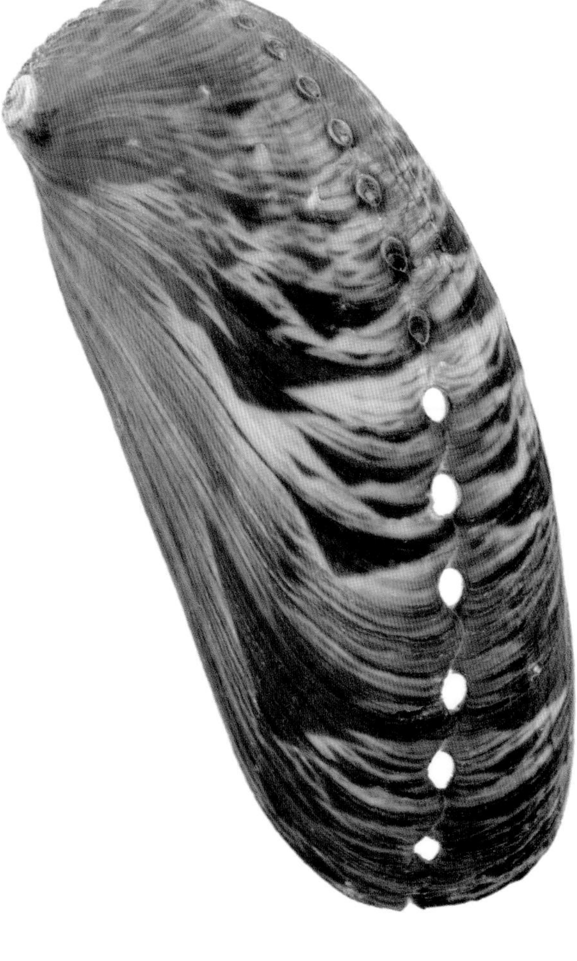

Green tree snail

Papustyla pulcherrima

The green tree snails of Manus Island in the South Pacific Ocean are one of only a few molluscs that have shells the colour of vivid leaf-green. These land snails live high in the forest canopy where they feed on detritus or fungi and lichens that grow on the trees. Once the shells became popular internationally in the 1930s, demand for them grew, particularly for jewellery, and, as a result, the species is in danger of extinction. They are also endangered because of the destruction of their habitat through forest clearance. Conservation efforts have increased in recent years and now the snail and its shell are protected under the Convention on International Trade in Endangered Species (CITES). This agreement aims to ensure that trade in specimens of wild animals and plants does not threaten their survival. The green tree snail was also the first invertebrate to be listed on the Endangered Species Act of the United States of America in 1973.

Distribution
Manus Island, Admiralty Islands

Size
35 mm (1½ in)

Bean partula

Partula faba

This species of tree snail was endemic to the islands of Raiatea and Taha'a in French Polynesia. It is now extinct due to the introduction in the late 1980s of the small carnivorous rosy wolfsnail, *Euglandina rosea*, which caused the decline of many native species of Partulidae (land snails) including *Partula faba*. Ironically, the rosy wolfsnail was introduced to control the population of the giant African land snails, another introduced species that was presenting a threat to the local flora. In an effort to save the species from extinction, conservationists rescued a group of the bean partula snails from the island in 1991 to breed in captivity. First London Zoo, then Bristol Zoo and lastly Edinburgh Zoo were entrusted with the last-known colony of these snails. Unfortunately, the captive snails failed to reproduce and the population declined slowly. The last known bean partula snail sadly died on 21 February 2016.

Distribution
Extinct

Size
25 mm (1 in)

Sunburst star turban

Astraea heliotropium

The sunburst star turban was one of the most attractive shells brought to Europe by the officers and crew who sailed around the world with Captain James Cook. James Cook (1728–1779) was a British explorer and captain in the Royal Navy. In 1768 he was commissioned to command a scientific voyage to the Pacific Ocean, and from 1768 to 1771 he led his first voyage to the South Seas on HMS *Endeavour.* The crew were the first Europeans to see these shells when they arrived in New Zealand during Cook's voyages to the South Seas. This species lives on rocky reefs and is herbivorous, grazing on algae.

Distribution
New Zealand

Size
75 mm (3 in)

Money cowry

Monetaria moneta

Throughout the world, and up until the end of the Roman Empire in AD 500, this little shell was used as currency. It continued to be used as small change in the markets of Benin, Ghana, Ivory Coast and Burkina Faso during the 20th century. Many factors contributed to the shells' success as money, but probably the most significant is that they could not easily be counterfeited (however, imitations were produced from bones, ivory, stones and bronze), they were a convenient and constant size and, like coins, they were durable and could be transported easily. In the 18th century in Uganda, one could purchase a cow for 2,500 of these shells, a goat for 500 and a chicken for 25 shells.

Distribution
Indo-Pacific

Size
15 mm (½ in)

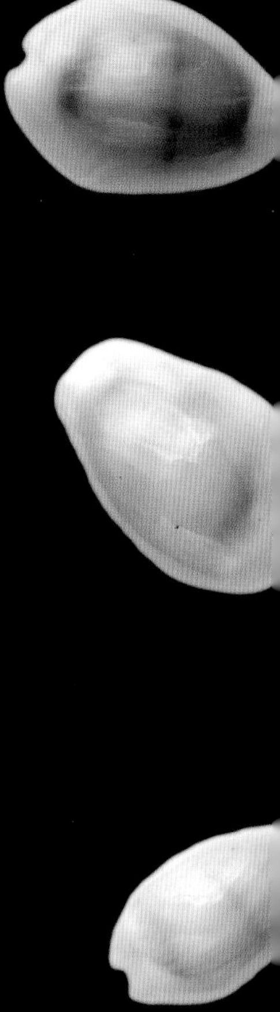

West Indian worm shell

Vermicularia spirata

Most gastropods have regular and neatly coiled shells but the West Indian worm shell is an exception. Its shell is thin, twisted and tube-like. The first whorls curl regularly, but as the animal grows, the later whorls become progressively more uncoiled and distorted, so that no two specimens of the same species are identical, but they do look alike. These marine snails are sequential hermaphrodites - they start their adult life as free-living males, but later change sex to become females, though it is not known why they do this. At this stage, they become sessile organisms, attaching themselves to various substrates such as sponges and other colonial animals.

Distribution
Northwestern Atlantic Ocean, Caribbean Sea and the Gulf of Mexico

Size
100 mm (4 in)

Cuban land snail

Polymita picta

Cuba is home to over 1,400 species of land snails, of which about 96 per cent are not found anywhere else in the world. Its complex geology and varied topography have allowed a diversity of ecosystems and species to evolve, creating one of the most biologically diverse tropical islands on Earth, and one of the most spectacular malacofaunas on the planet. The most famous Cuban land snails are those of the genus *Polymita*, known also as painted snails because of their variety of vibrant colours. They can be green, red, yellow, black, brown or a combination of colours, but never blue. They live on trees and bushes feeding on lichen, moss and fungi, and are preyed upon by an array of vertebrates and invertebrates, but habitat destruction is by far their greatest threat.

Distribution
Endemic to Cuba

Size
25 mm (1 in)

Ventral harp

Harpa cabriti

Harp snails have the ability to break off a piece of their foot when threatened by a predator or when disturbed in some way, much like a lizard may shed part of its tail when attempting to escape. The detached piece of tissue that is left behind continues to wriggle, to distract the predator long enough for the harp shell to burrow to safety under the sand. The capacity for self-amputation is called autotomy – it does no harm to the animal and the harp snail is able to regenerate the missing part, so it can use the same strategy again.

Distribution
Red Sea, Arabian Gulf and Indian Ocean

Size
100 mm (4 in)

Cambodian bertia

Bertia cambojiensis

Most land snails are hermaphrodites, meaning that they have both male and female reproductive organs and are capable of both transferring sperm and laying eggs. Reproduction is usually sexual, although self-fertilization may also occur in some species. Some land snails, such as the Cambodian bertia snail, perform courtship behaviours before mating, during which they shoot one or more love darts into each other's bodies. The function of the dart in mating has been the subject of much study, and new discoveries are still being made. The dart is made of calcium carbonate, chitin or cartilage, and can be from 1 to 30 mm (0.04 to 1¼ in) in length. The dart is formed and stored internally in a dart sac and is shaped differently depending on the species.

Distribution
Vietnam

Size
60 mm (2½ in)

Dennison's morum

Morum dennisoni

Shells have appeared for sale in auctions since the late 17th century. Many museums, in addition to regular donations, increased their collections through purchases from dealers and auction rooms. One of the most notable auction houses in London was Stevens' Auction Rooms, based at 38 King Street, Covent Garden, which specialized in the sale of natural history specimens and ethnographic artefacts. For almost two centuries some of the most famous shell auctions took place at this address, providing specimens to museums, collectors and scientists. The Dennison's morum sea snail is a Caribbean rarity first described from John Dennison's (d. ?1864) shell collection. It is a carnivorous gastropod that lives on sandy and muddy bottoms, and it is usually found in deep waters. The Dennison collection was the most celebrated of the 19th century, and its auction sale in 1865 was considered by collectors of natural history to be the sale of the century.

Distribution
Caribbean Sea

Size
50 mm (2 in)

Atlantic fig shell

Ficus papyratia

The Academy of Natural Sciences of Drexel University, formerly the Academy of Natural Sciences of Philadelphia, was founded in 1812 by a group of amateur scientists with an interest in natural history. It is the oldest natural science research institution and museum in the Americas and houses the oldest shell collection in the USA. Thomas Say (1787–1834), the first mollusc curator of the Academy and one of its founding members, discovered this mollusc during a journey to East Florida, and described it in 1822. The shell is shaped like a fig, hence the common name.

Distribution
North Carolina to the Gulf of Mexico

Size
85 mm (3 ¼ in)

Tapestry turban

Turbo petholatus

Most gastropods can close the entrance to their shells with a kind of trap door attached to the back of their foot. This is called the operculum, meaning little lid, and it is shaped to exactly fit the shell opening, protecting the animal against predators and desiccation. This structure can be corneous (made of conchiolin) or calcareous (made of calcium carbonate) and is present in many groups of sea snails, freshwater snails and in some land snails. The turban shells are a large group of marine snails mostly found in the tropics in shallow water, often near coral reefs grazing on algae. They have a calcareous and heavy operculum. The tapestry turban's operculum is known as a cat's eye; it is brightly coloured, and is often used for making jewellery.

Distribution
Red Sea, Indo-Pacific

Size
70 mm (2 ¾ in)

Saul's murex

Chicoreus saulii

Muricidae are a widely distributed group of marine snails comprising hundreds of species that vary in shape and size. They are voracious predators, some of them feeding on other molluscs by drilling a small circular hole in the shell of their prey, using their radula. They also feed on corals, barnacles and echinoderms. Their habitats vary, with some species living among rocks and others in muddy or sandy places. The female of the species lays egg capsules and these are often found under rocks. On many occasions they congregate together, particularly during the early summer, to deposit their eggs in one large mass. The shells of this group are the most complexly sculptured of all the seashells. The most important feature is the numerous varices, which often have prominent and sometimes bizarre spines, nodules or fronds. An example of the fronded species is Saul's murex, a shell with distinctive branching fronds, occasionally with pink tips.

Distribution
Southern Pacific

Size
110 mm (4 ¼ in)

Cuming's periwinkle

Tectarius cumingii

Hugh Cuming (1791–1865) was an English collector and a pioneer in a new era in the story of conchology. He collected tens of thousands of shells, many of them new to science, on expeditions to the Pacific coast of the Americas and the Indo-Pacific. His vast collections became famous among naturalists and inspired some of the pivotal works of conchological literature. The Cuming Collection, the largest private collection of molluscs of his day comprising almost 83,000 specimens, was purchased by the British Museum in 1866. The great value of this collection was in the large number of original specimens that it contained and the diversity of taxa. It remains as relevant today, supporting a vast amount of multidisciplinary research globally. This sculptured periwinkle was named after Cuming in 1846 by the German-Chilean malacologist R.A. Philippi (1808–1904).

Distribution
Central Indo-Pacific

Size
20 mm (¾ in)

Annam amphidromus

Amphidromus inversus

The direction in which the whorls are coiled is an important shell characteristic. The majority of shells are coiled to the right or clockwise (dextral) but a few species coil to the left or counterclockwise (sinistral). However, the amphidromus tree snails are unusual; they are chirally dimorphic, meaning that in the same population you can find both dextral and sinistral individuals. One of the reasons for this exception could be related to their natural predators, snakes. Most snail-eating snakes of the genus *Pareas* have elongated teeth on the right mandible, and for that reason specialize in predation on the dextral individuals. The snails have evolved by coiling the other way, which makes it difficult for the snake to grasp and eat them.

Distribution
Southeast Asia

Size
45 mm (1¾ in)

Pelican's foot

Aporrhais pespelecani

The Conchological Society of Great Britain & Ireland is one of the oldest existing natural history societies devoted to the study of molluscs. It was founded in 1876 as the Leeds Conchological Club, changing its name at the 39th meeting of the society, held on 30 May 1878. The society owes its origin to four enthusiastic conchologists, William Nelson (1835–1906) who was the first president, John William Taylor (1845–1931), William Denison Roebuck (1851–1919) and Henry Crowther (1848–1937). The society's logo is the pelican's foot shell, a European species of sea snail that lives on sandy or muddy bottoms. The snail's common name comes from the shape of the adult shell that resembles the webbed foot of the pelican.

Distribution
Northern Norway and Iceland to the Mediterranean

Size
40 mm (1½ in)

Juke's keyhole limpet

Diodora jukesii

Keyhole limpets are a group of marine snails, named after the small hole at the top of the shell through which the waste waters are expelled. Water enters under the shell's lip, passes upwards over the gills, and exits through the hole, collecting waste products as it flows out. This way, faeces are swept well clear of the animal and the inhaled water current does not become contaminated by them. The shape of the opening is variable, and it appears in various positions depending on the species. Keyhole limpets occur worldwide, mostly living in shallow water or in the intertidal zone where some of them feed on algae, others on sponges, and some consume detritus. The Juke's keyhole limpet can be found living under rocks on intertidal reefs in southern Australia.

Distribution
Southern Australia

Size
25 mm (1 in)

South African turban

Turbo sarmaticus

The South African turban is a species of sea snail with a large, thick and heavy shell. It is one of the most appreciated and sought after by shell enthusiasts, perhaps due to its aesthetic appeal. In its natural state the shell is not very attractive. However, when its outer surface is removed by acid or abrasion, it reveals a beautiful mother-of-pearl layer underneath with elaborate, banded colour patterns. The silvery mother-of-pearl combined with its natural, rusty-brown colour makes it the perfect material for jewellery. It is native to South Africa where the animal is an important food item. Archaeological records of shell middens also show that the species has been harvested for food since prehistoric times.

Distribution
Endemic to South Africa

Size
60 mm (2 ½ in)

Zebra volute

Amoria zebra

Volutes are a large group of sea snails that are popular with collectors because of the variety of colours and patterns on their smooth shells. They are mainly tropical in distribution, with the highest diversity around Australia. Most volutes inhabit the sandy or muddy bottoms of the continental shelf and slope. They are carnivorous, feeding on small animals, including other molluscs, which they locate by smell. The prey is enveloped and suffocated by the volute's large and inflated foot. This often very colourful foot also enables them to burrow in sand and crawl quicker than most other marine snails. The zebra volute is endemic to Australia, ranging from Queensland to New South Wales, where it is found in large colonies. Its shell is white to light brown, with distinctive brown lines.

Distribution
Australia

Size
40 mm (1½ in)

Crowned baler

Melo aethiopicus

Sir Hans Sloane (1660–1753) was an Irish physician, scientist and collector. During his life, he amassed one of the greatest private collections of plants, animals, antiquities, coins, books, drawings and maps, as well as precious antiques and curiosities from all over the world. The outstanding natural history collection was in part due to a far-reaching network of traders, apothecaries, physicians, naturalists and collectors, who supplied Sloane with specimens from distant lands. After his death, the UK government acquired the Sloane collection of more than 400,000 objects for £20,000 (more than £4 million today), including around 6,000 shells. Today, around 700 of the shells are still housed in the mollusca section of the Natural History Museum, London, including a specimen of a crowned baler. These large and rounded shells are often used in Indonesia to carry water, or for bailing out water from canoes - hence their common name.

Distribution
Indonesia

Size
240 mm (9 ½ in)

Australian trumpet

Syrinx aruanus

The Australian trumpet is the world's largest living shelled gastropod, growing up to 910 mm (36 in) in length, and it is also the heaviest, weighing up to 18 kg (40 lb). It lives on muddy seabeds up to 40 m (130 ft) in depth and is an extremely active predator, feeding on large polychaete worms. This gastropod is fished for its meat, and large ones are a popular target of shell collectors and are prized as a decorative object. For this reason, populations in easily accessible habitats have declined, causing concern about its conservation status. The shell is also used to carry water and as a trumpet.

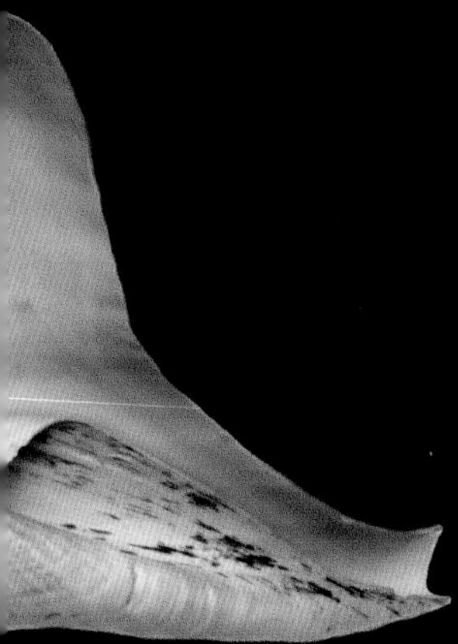

Distribution
Northern Australia, Papua New Guinea

Size
600 mm (23 ½ in)

Spotted tun

Tonna dolium

The tuns are a group with thin, inflated shells and a very wide aperture. There are about 30 living species in the family Tonnidae worldwide, confined to continental shelves and slopes in warm-temperate and tropical seas. They are sand-dwelling, nocturnal species, common at depths of about 10 m (33 ft) or more, and are carnivorous, eating mainly sea cucumbers. They first paralyze their prey with a salivary secretion that contains sulphuric acid and then swallow them whole. The ingestion of such large prey is the reason for the wide, inflated shell.

Distribution
Indo-Pacific

Size
80 mm (3 in)

Spiny riversnail

Io fluvialis

The American Malacological Union, now the American Malacological Society, was founded in 1931 mainly through the organizing efforts of Norman W. Lermond (1861–1944), a New England amateur naturalist. It is one of the oldest active societies of malacology in the USA. Its first meeting was held on 30 April in the same year at the Academy of Natural Sciences in Philadelphia, with 29 attendees from 12 states. The first president was Henry A. Pilsbry (1862–1957) the curator of the Department of Shells. Since 1960 the Society's logo has been the spiny riversnail, a species of freshwater snail that is endemic to the Tennessee River and its larger tributaries. This species inhabits rocky riffles with good flow, feeding on the algae coating them, but it is now threatened by habitat alteration due to the impoundment of the Tennessee River and declining water quality.

Distribution
Tennessee River and its larger tributaries (USA)

Size
50 mm (2 in)

Geography cone

Conus geographus

Cone snails move slowly but they have an adaptation that enables them to hunt - they can shoot a harpoon-like radula that paralyzes their prey almost instantly. The radula, or modified tooth, delivers conotoxin, a venom so powerful that it can kill humans, and the reason why the geography cone is considered to be one of the deadliest animals in the world. Envenomation in humans by *Conus geographus* is extremely rare, with only thirty or so deaths attributed to this animal in 300 years. Nevertheless, we have not managed to create an antivenom for a cone snail sting, so treatment is limited to keeping victims alive until the toxins wear off. However, research shows that the venom has great potential in medicine as certain components of the venom can be up to 10,000 times more potent than morphine.

Distribution
Indo-Pacific

Size
90 mm (3 ½ in)

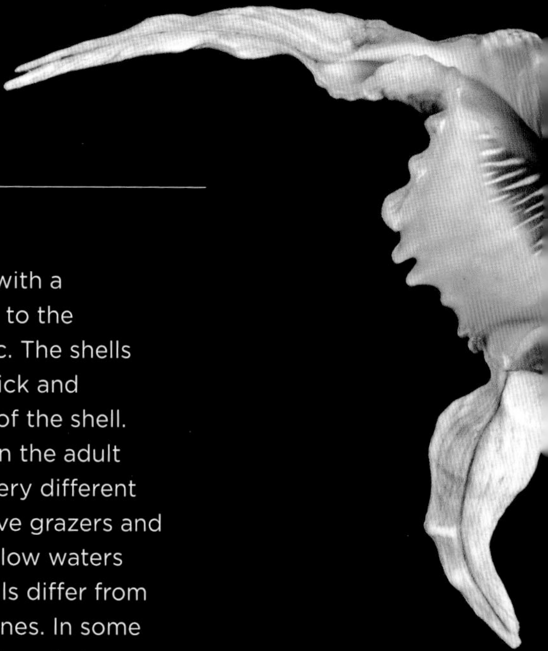

Scorpio conch

Lambis scorpius

Spider conchs are marine snails with a geographical distribution limited to the warmest parts of the Indo-Pacific. The shells are characterized by the long, thick and curved projections along the lip of the shell. These features are only present in the adult shell, so young shells resemble very different gastropods. The animals are active grazers and their preferred habitat is the shallow waters around coral reefs. The male shells differ from females in size, sculpture and spines. In some species the male may be as much as forty-five per cent smaller than the female. This difference of appearance between male and female of the same species is called sexual dimorphism. The scorpio conch is one of the longer-spined species, with a bright, purple-coloured aperture that also displays a series of white stripes.

Distribution
Indo-Pacific

Size
70 mm (2 ¾ in)

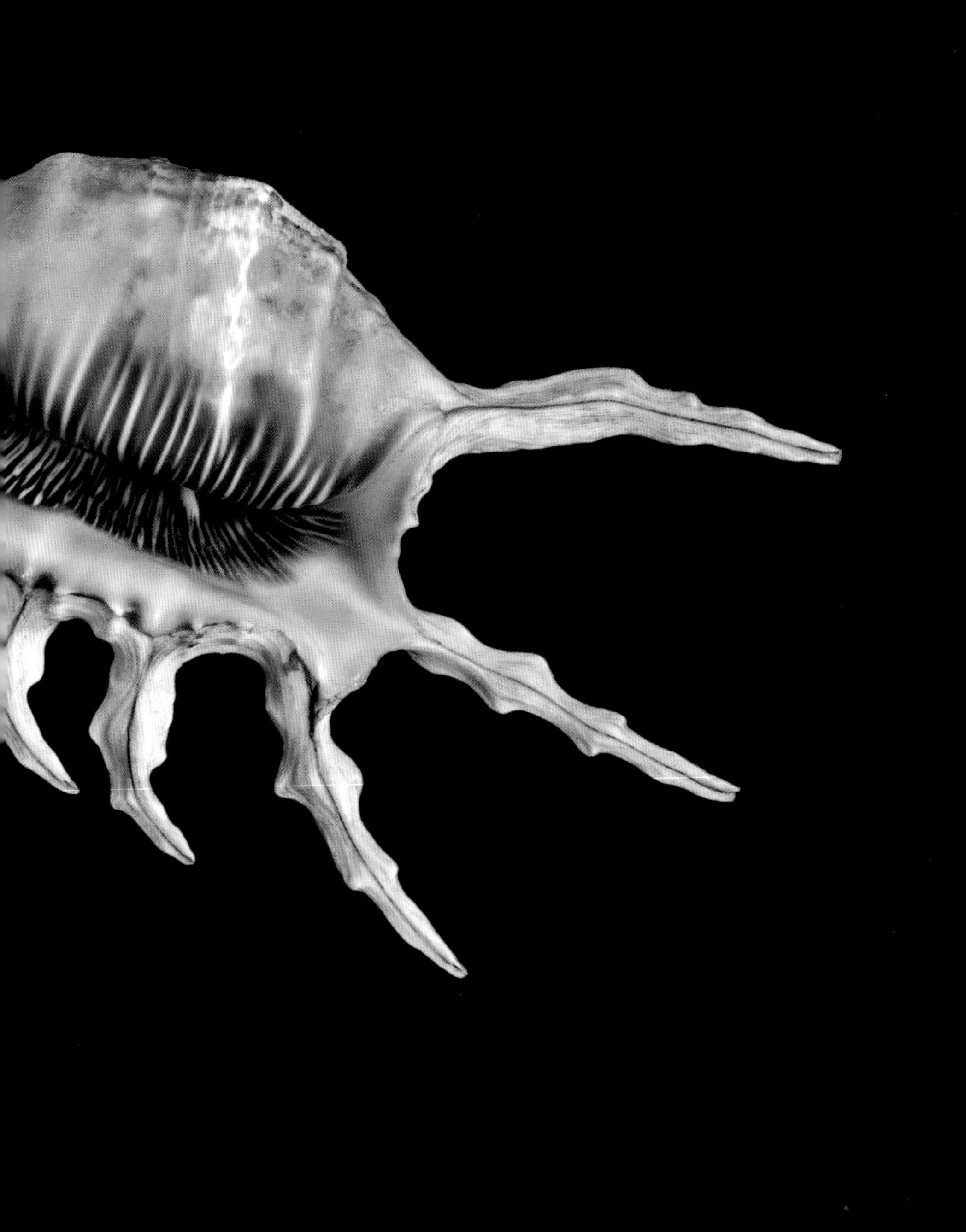

Black-tankard angaria

Angaria melanacantha

Conchologia iconica, or Illustrations of the Shells of Molluscous Animals, is one of the greatest classical works of conchology. The publication, authored by Lovell Augustus Reeve (1814–1865), was begun in January 1843 and finished in 1878, reaching 20 volumes, containing 281 monographs and about 27,000 drawings of shells at life-size. After Reeve's death, George Brettingham Sowerby II (1812–1884), a British illustrator and a conchologist, completed the work. Sowerby also drew and lithographed almost all the plates for the entire work. Reeve's aim was to illustrate all species of shells, whether previously described or not. This angaria shell is an example of a species he figured in the first volume. It is a species of sea snail with distinct rows of black scale-like spines that bend inwards at the top. It lives in deep water near coral reefs, and is herbivorous, grazing on algae.

Distribution
Philippines

Size
40 mm (1½ in)

Hirase's volute

Fulgoraria hirasei

Yoichirō Hirase (1859–1925) was a Japanese malacologist and shell dealer who contributed to the advance of malacology in Japan. He coordinated numerous field trips, discovering and naming many new species in collaboration with scientists from around the world. He assembled the largest collection of shells in Japan and had his own shell museum in Kyoto from 1913 to 1919. His most famous book is *Kai Chigusa*, more commonly known as *Kai sen shu* (*The Illustrations of a Thousand Shells*), which was produced from 1914 to 1922. There are four volumes containing 400 illustrations constructed in a traditional technique known as *orihon* (a long strip of paper with writing on only one side, and then concertina-folded). This beautiful shell, a marine predator endemic to Japan, was named after him in recognition of his work.

Distribution
Japan

Size
125 mm (5 in)

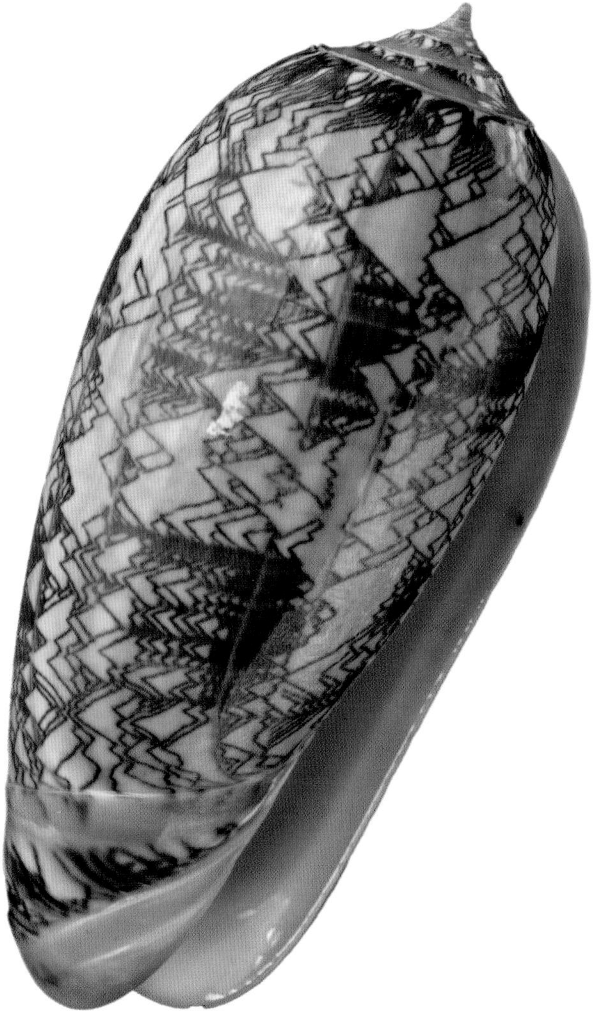

Tent olive

Oliva porphyria

Olive shells have always been popular with collectors, as they are glossy, smooth, very solid and exhibit a wide variety of colours and patterns. They are a large group of marine snails that inhabit all tropical and warm seas globally. Olives live in colonies buried in sand and are scavengers as well as active predators, feeding on small mussels, gastropods, crabs and carrion. The tent olive is the largest of all the olives, with markings reminiscent of a large encampment of various sized tents.

Distribution
Gulf of California to Panama

Size
110 mm (4 ¼ in)

Spiral babylon

Babylonia spirata

This beautiful marine snail is easily recognized by its smooth, often polished shell, and by the distinctive patterning of orange on white. The spire, with its almost vertical whorls and deep channels between them, resembles popular illustrations of the mythical Tower of Babel, a biblical building with stepped sides; this similarity gave its name to the genus. It is primarily a scavenger feeding on dead fish and usually lives in shallow, subtidal waters but can also be found in deeper waters, on sandy and muddy bottoms. Several species of babylon snails are eaten in Asia, where they are caught with bamboo baskets baited with dead fish. Nothing is wasted and after the animal is eaten, the shell is sold as a souvenir.

Distribution
Indian Ocean

Size
65 mm (2 ½ in)

Flat periwinkle

Littorina obtusata

The lives of animals on rocky shores seems to be dominated by challenging factors including desiccation, wave action, deluges of rainwater and extremes of temperature. Despite this, there is often a diverse and abundant molluscan fauna living between the tidal levels. Among the best-known are the periwinkles, which can be found in huge numbers, grazing on fine algae. Their shell may be smooth and thick and range in colour from white and yellow to violet, green and brown. They safely hide under seaweed, in holes, crevices and in disused barnacle shells. The flat periwinkle is a very common species and is usually found among the large seaweeds, especially black wrack, where the animal is kept moist and shaded from the sun when the tide is out.

Distribution
From southern Labrador to New Jersey; Norway to the Mediterranean

Size
15 mm (½ in)

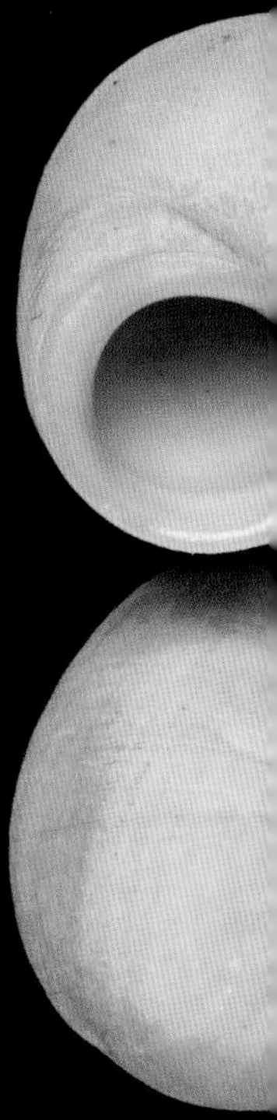

Marbled cone

Conus marmoreus

One hundred years before being described by Carl Linnaeus, the marbled cone had already been etched by the famous Dutch painter Rembrandt van Rijn (1606–1669) in his 1650 work entitled *The Shell*. The etching was based on a shell he had in his small natural history collection. He painted it life-size, mastering the shape and pattern of the shell, but he mistakenly showed it as being left-handed or sinistral. His shell displays an anticlockwise twist, which does not occur in nature. Snail shells of this kind always coil in a right-handed or clockwise direction.

Distribution
Indo-Pacific

Size
80 mm (3 in)

Pyramid clio

Clio pyramidata

This small, fragile and translucent shell belongs to a group of pelagic gastropods called pteropods. The world's oceans are heavily populated with these small snails, which form part of the plankton. They provide a vital source of food for many small and large aquatic organisms. They are also named sea butterflies because the animal has two large wing-like extensions on its foot, which can be flapped like the wings of a butterfly. Since their shells are exceptionally vulnerable to rising levels of carbon dioxide in the oceans, these organisms have been proposed as bioindicators to monitor the effects of ocean acidification. This process occurs when oceanic waters absorb the atmospheric carbon dioxide released mainly from the burning of fossil fuels, making the water more acidic. At a certain threshold, the acidity dissolves the snails' shells, and pteropods, like other molluscs, cannot survive without their shell to protect them.

Distribution
Worldwide, open seas

Size
10 mm (½ in)

Venus comb

Murex pecten

The venus comb shell is one of the most elegant of the mollusc shells. It is covered in more than 100 needle-sharp and evenly spaced delicate spines. The spines are formed by projections of the mantle that secrete calcium carbonate. Amazingly the mantle can also reabsorb the spines of the preceding whorl, so the animal and shell can carry on growing without the spines from the previous whorl blocking the aperture. The function of the spines is still not fully understood, but they might act like a cage to entrap mobile prey, to keep the shell from sinking or overturning on soft sediment, or as a defence against predators.

Distribution
Indo-Pacific

Size
120 mm (4 ¾ in)

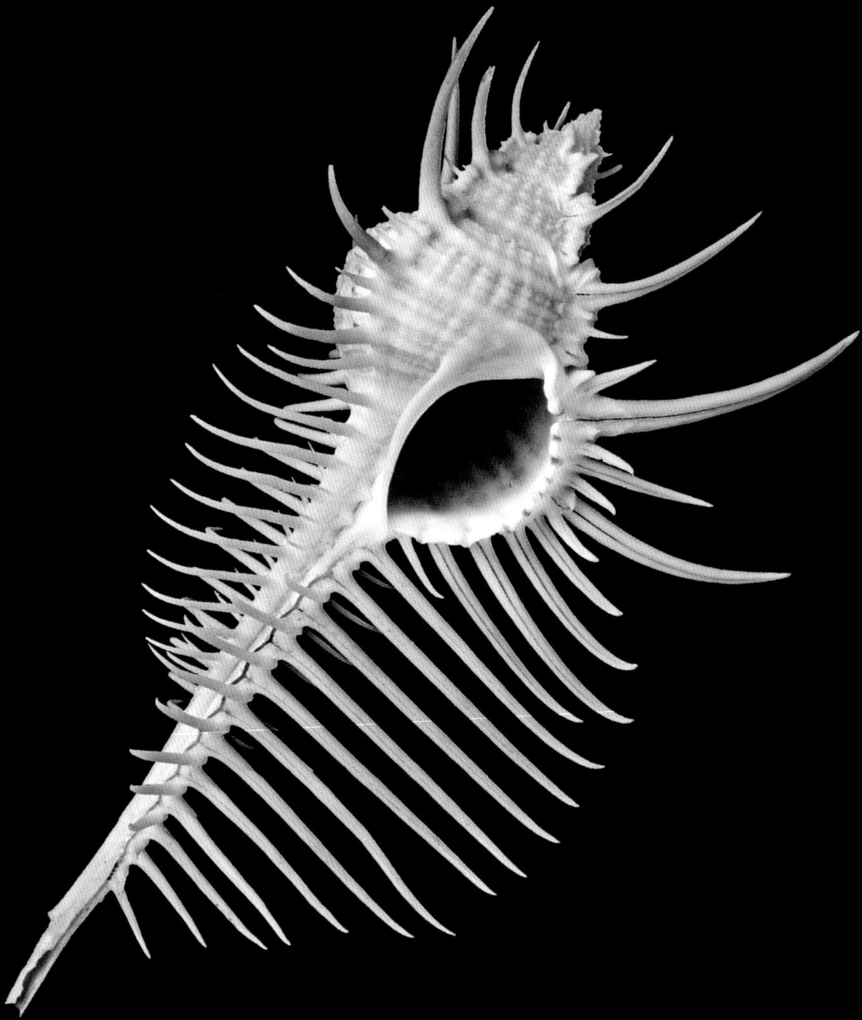

Jenner's or pustulate cowry

Jenneria pustulata

This unusual shell is similar in shape to what is known as a 'true cowry' but it lacks a true cowry's smooth surface. Instead, the exterior is covered with numerous bright orange nodules or pustules (hence the species name *pustulata*), surrounded by a dark ring and with a ridged base. The coloration and texture of the shell and mantle blend with the encrusted rocks on which these coral-eating animals live.

Distribution
Southern California to Panama

Size
20 mm (¾ in)

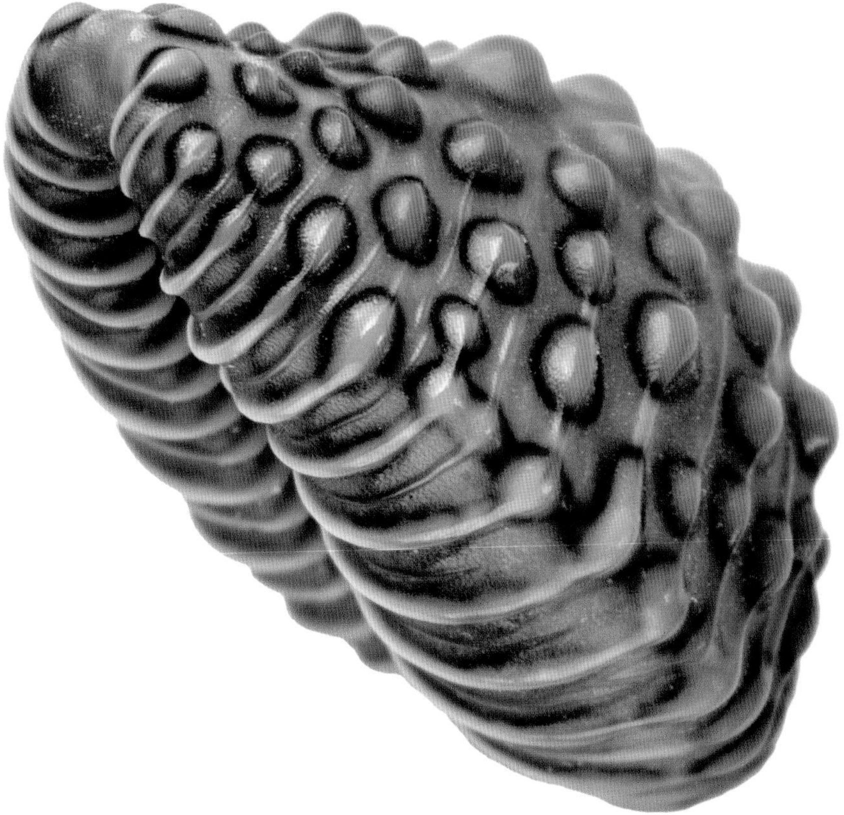

Arabic bubble shell

Bulla arabica

Bubble shells have, as the name suggests, a bubble-shaped shell, into which the animal is able to withdraw completely. They are a worldwide group of marine gastropods. The animal shelters by burrowing in sand or mud, often among marine grasses, in protected bays and lagoons. They are nocturnal and herbivorous animals, emerging at night to feed mainly on small seaweed and grasses. The bubble shells belong to a vast group of gastropods that also include the canoe shells, the colourful sea slugs and the sea hares. The Arabic bubble shell has a large, thick, brown mottled shell and only occurs in the Red Sea and in the Arabian Peninsula.

Distribution
Red Sea and Arabian Peninsula

Size
30 mm (1¼ in)

China limpet

Patella ulyssiponensis

Limpets are familiar inhabitants on rocky shores throughout the world's oceans from tropical to polar regions. Unlike most other gastropods, the shell is a simple cone and is not coiled. The cone is an ideal shape to resist the force of breaking waves. When the tide goes out, the muscular foot of the limpet pulls the shell tightly against the rock, which prevents dehydration of the animal when it is out of water. At high tide, they move around grazing on algae, which they scrape from the rocks using their radula. Many species excavate a shallow home-scar on the rock where they live, which exactly fits the contour of the shell. They leave the scar to feed but return to the same spot at every low tide. This behaviour is called homing, and it is thought that limpets can do this because they follow pheromones secreted in the outgoing mucus trail left behind by their foot.

Distribution
From the British Isles to the Mediterranean

Size
40 mm (1½ in)

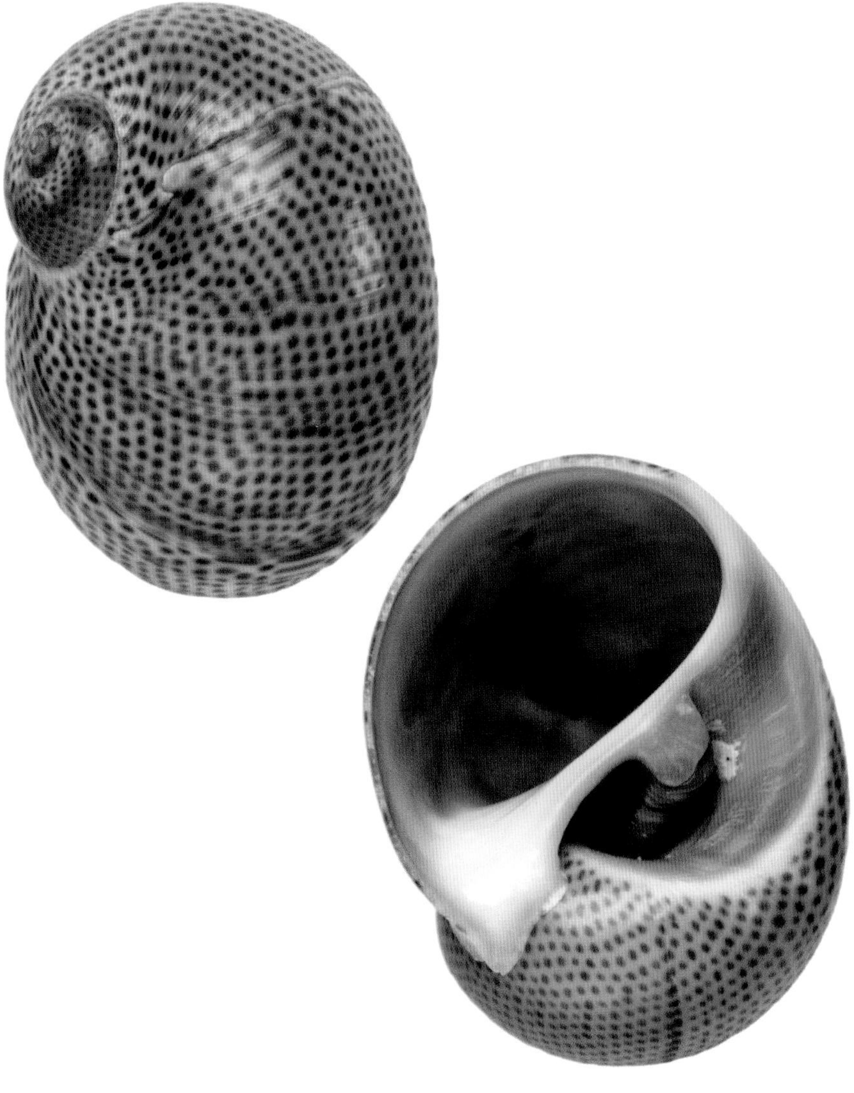

Maple leaf triton

Gyrineum perca

The maple leaf triton is a species of marine snail that has a shell with a distinctive outline making it easily recognizable. It looks like it has been squashed flat, with remarkable long wing-like varices fanning out on either side like a maple leaf. Snails such as this, which live on the surface of a soft substrate like sandy or muddy bottoms, tend to reinforce their shells by periodically thickening or flaring the shell edge to form a varix. The snail grows in spurts, adding shell material rapidly and in fixed increments, after which it rests for several weeks and secretes a flattened varix. The resulting shell shape is a beautiful and practical adaptation that helps to prevent the shell from sinking into the soft substrate on which it lives.

Distribution
Indo-Pacific

Size
70 mm (2 ¾ in)

Senckenberg false hadra

Euhadra senckenbergiana

The journal *Archiv für Molluskenkunde* was founded in 1868 as *Nachrichtsblatt der Deutschen Malakozoologischen Gesellschaft* (Newsletter of the German Malacological Society). It has been published continuously since then, making it the oldest malacological journal in the world. The first editor of this international journal was Wilhelm Kobelt (1840–1916), a German zoologist and curator of the Senckenberg Research Institute and Natural History Museum in Frankfurt, Germany. It is still edited by the curators of the mollusca department at this institution. This shell is a species of air-breathing land snail only found in Japan. It was described by Kobelt in 1875 and named in honour of the Senckenberg Museum.

Distribution
Japan

Size
55 mm (2 ¼ in)

Boxall's snorkel snail

Rhiostoma boxalli

Henry Haversham Godwin-Austen (1834–1923) was an English topographer, geologist, surveyor (K2, the second highest mountain in the world was once named Mount Godwin-Austen in his honour), explorer and a leading authority on Indian mollusca. During his surveying expeditions he collected molluscs from little known corners of India, which he systematically studied during his retirement. He was the author of *The Land and Freshwater Mollusca of India*, a work that brought him global recognition as a malacologist. Much of his collection is now in the Natural History Museum, London. This land snail was one of the many species described by Godwin-Austen. It is a species of snorkel snail, named due to a small tube found on the final whorl of its shell, which resembles a snorkel. This structure allows the snail to breathe even if its operculum, which covers the opening of the shell, is closed.

Distribution
Northern Borneo, Palawan Island

Size
25 mm (1 in)

Eloise’s acteon

Punctacteon eloiseae

Eloise’s acteon is the most beautiful shell in a group of predatory sea snails known as barrel bubble snails. It is endemic to the island of Masirah in Oman and is found only on a few muddy sand beaches – Eloise Beach is named as such due to the presence of this gastropod. However, in recent years it has become very difficult to find living examples and local biologists consider it to be over-collected and vulnerable. It was discovered by Donald Bosch (1917–2012) and named after his wife and collecting companion Eloise (1919–2016), and quickly became the iconic shell of Oman. The Bosch family were responsible for collecting several new species inhabiting the shallow waters around the shores of the Sultanate of Oman.

Distribution
Oman

Size
30 mm (1¼ in)

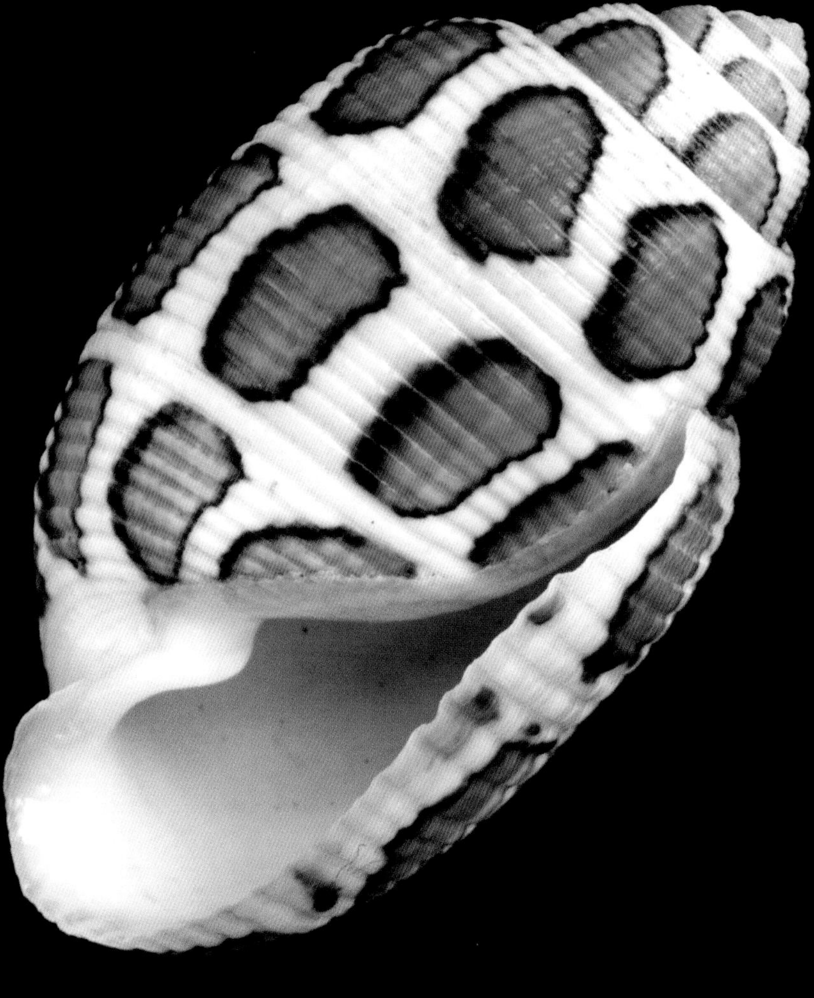

Bullmouth helmet

Cypraecassis rufa

Cameo carving is a craft for which only certain shells are suitable, like the large helmet shells. The inner and outer layers of these gastropods' shells are of different colours, allowing craftsmen to take decorative advantage of the contrast between them to make beautiful intricate cameos. For centuries cameo-makers have sculpted faces and figures and other exquisite designs from the outer layer, while the colourful inner layer serves as a contrasting background. The cameo first made its appearance in Alexandria, Egypt in 332 BC, but only found widespread popularity during the neoclassical revival in Europe of the 18th and 19th centuries, when they also became fashionable items of jewellery.

Distribution
Indo-Pacific

Size
95 mm (3 ¾ in)

Scaly-foot snail or sea pangolin

Chrysomallon squamiferum

The scaly-foot snail was first discovered in 2001 at Kairei hydrothermal vent field, on the Central Indian Ridge (the north–south mid-oceanic ridge in the western Indian Ocean), and later found in four other vent fields. It lives at depths of between 2,400 and 2,800 m (7,875–9,185 ft), on the edge of hydrothermal vents and black smokers that can reach temperatures of 300–400°C (572–752°F). The foot of this snail is unique among gastropods because it is covered in hundreds of iron-infused scales. These scales are fleshy in the centre and hard on the exterior. Research suggests that the primary function of these scales is detoxification of sulphur metabolites, originating from symbiotic bacteria the snail cultivates inside its cells for nutrition. Iron sulphide is also found in the shell. In 2019, the International Union for Conservation of Nature (IUCN) listed this extraordinary snail as Endangered, the first species to be assigned this status due to risks from deep-sea mining.

Distribution
Indian Ocean

Size
35 mm (1½ in)

Lined moon snail

Tanea lineata

For some molluscs, like most bivalves, chitons and tusk shells, the duties of reproduction are finished once their eggs have been cast unprotected into the water. Others, like some cephalopods, take care of their fertilized eggs and sometimes their young. Marine snails lay their eggs within gelatinous or leathery capsules in a protected place, often attached to rocks, stone or rubble. The moon snails adopt an unusual method of depositing their eggs. They make a distinctively shaped egg mass at night – a necklace or collar-like ribbon made of sand grains cemented with mucus. The fertilized eggs are then incorporated into the inner side. These coiled egg ribbons may often be seen lying on sandy beaches. Each species of moon snail produces its own characteristically shaped egg mass.

Distribution
Japan to Queensland, northern Indian Ocean

Size
25 mm (1 in)

Loebbecke's murex

Chicoreus loebbeckei

Loebbecke's murex is a marine gastropod with three beautiful wing-like frilled varices on its shell. Its typical colour is orange, but it is very variable, ranging from white to pink to yellow to red – white is the rarest form. It is a very rare species in collections due to the fact that its lives in deep water and is therefore difficult to collect, especially with its fragile varices still intact. This species was described in 1879 by Wilhelm Kobelt (1840–1916), a German malacologist. The shell belonged to the private collection of the German pharmacist Theodor Löbbecke (1821–1901), whom Kobelt named it after.

Distribution
Indo-Pacific

Size
60 mm (2 ½ in)

Horned helmet

Cassis cornuta

Coral reefs are one of the richest and most diverse environments on Earth, providing complex and varied marine habitats that support a wide range of organisms. They cover less than one per cent of the global marine ecosystem, yet they provide a home for 25 per cent of all marine species including thousands of mollusc species. Coral reefs are under threat worldwide. In addition to anthropogenic challenges, one of the major concerns is the crown-of-thorns starfish. These starfishes can grow as large as 1 m (3 ft) in diameter, and they feed on coral. The horned helmet is one of the few species of mollusc that predates on the crown-of-thorns starfish. It is for this reason that it has been put under strict protection in Queensland, Australia, in a bid to prevent the starfish spreading through the Great Barrier Reef and consequently destroying it.

Distribution
Red Sea, Indo-Pacific

Size
270 mm (10 ½ in)

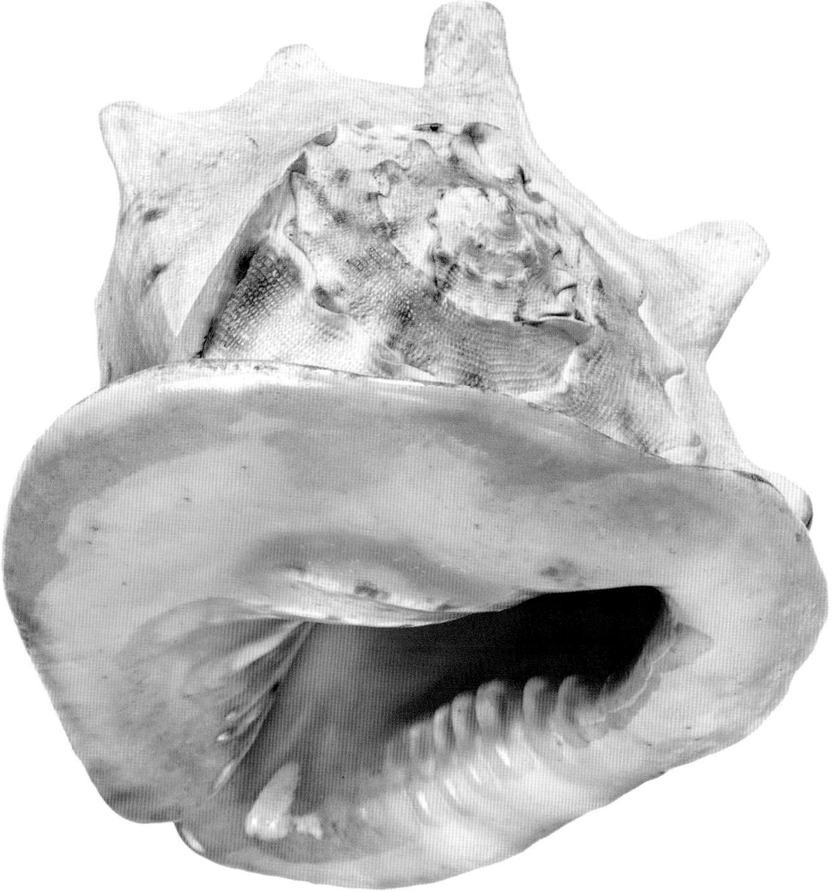

Babylon turrid

Turris babylonia

The turrids are the most diverse family of all seashells, with more than 4,000 species already described and an estimated 10,000 more still to be discovered and named. They live in sand, mud, rock and coral habitats, from the intertidal zone to the abyssal depths, and in all oceans including the Antarctic and the Arctic. All species are carnivorous, preying mainly on polychaete worms. Like cone shells, which they are closely related to, turrids and other toxoglossa have a venom gland and a harpoon-like radular tooth, which they use to hunt worms. Their shells also show a diversity of forms, but they all present a characteristic feature, a slit in the outer lip, known as the turrid notch, which is for the anal siphon, a fold of the animal tissue that directs waste away from the animal.

Distribution
Indo-Pacific

Size
85 mm (3 ¼ in)

Roissy's tree snail

Calocochlea roissyana

Land snails have adapted to an amazing range of environments and conditions, from snowy mountains to the steaming jungles of the tropics and from deep caves to the deserts of Africa. Some hibernate if it is too cold or survive in a state of suspended animation if it is too hot, sealing their aperture with an epiphragm – a layer of dried mucus. Land snails can also be found in a variety of substrates, in leaf litter, on rocks, under logs, up trees and on smaller plants. The greater the variety of microhabitats in a site, the greater the number of species you are likely to find. In the Philippines, there are hundreds of species of large, colourful tree snails, like the Roissy's tree snail, many of which have become extinct because of deforestation.

Distribution
Mindoro Province (Philippines)

Size
30 mm (1¼ in)

Textile nerite

Nerita textilis

Nerites are a group of gastropods that occupy an amazing variety of habitats. Most of the species live in colonies on rocky shores but others live in brackish estuaries, freshwater lakes and streams. Some species even live out of the water on trees at the base of the trunk and among the roots. They have a thick shell and a very close-fitting operculum – the animal can use this to trap water inside the shell, allowing it to survive out of water for extended periods. Some of them, for example this textile nerite, have even colonized rocky surfaces above the high tide that are only splashed by wave spray and never submerged. Nerites are herbivorous molluscs – they scrape algae off the rocks or trees where they live. The shell, with a few exceptions, has a globular shape. In the textile nerite the thick spiral ribs are broken by growth lines that give them the appearance of twisted cords.

Distribution
Eastern Africa to western Pacific

Size
35 mm (1½ in)

First pagoda shell

Columbarium pagoda

This spiralled marine shell belongs to a group of worldwide, deep-water molluscs commonly known as the pagoda shells, named after their resemblance to the tiered pagoda temples of Japan and other Asian countries. They inhabit sandy and muddy bottoms along the continental slope. The shell is very slender and spindle-shaped with a long siphonal canal. The spines are upturned, flattened and triangular in shape. The animal is carnivorous and uses a long proboscis to feed on tube-dwelling polychaetes (bristle worms).

Distribution
Japan to South China Sea

Size
75 mm (3 in)

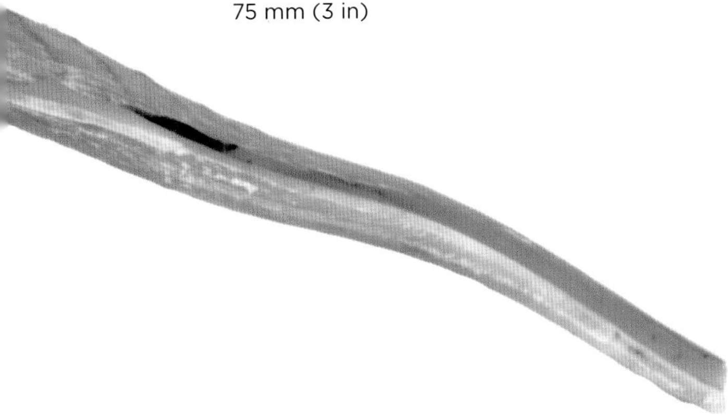

Loxostoma-oblique mouth door snail

Oospira loxostoma

William Henry Benson (1803–1870) was an Irish civil servant who worked for the East India Company in British India. When he first began collecting, the land snail fauna of India and the Andaman Islands were almost completely unknown. Benson pressed his family, household and friends to collect specimens, particularly land snails, and amassed a formidable collection of shells. It is for this reason that he is regarded as one of the pioneers of Indian malacology. His vast collection is scattered among many institutions and collections because the specimens were exchanged, donated or sold. This Indian land snail was described by Benson himself, and belongs to the clausiliids or door snails, a group of small, elongated, air-breathing gastropods.

Distribution
India

Size
25 mm (1 in)

Triseriate auger

Terebra triseriata

Augers are a group of marine snails with a long, fragile and slender shell, numerous whorls and a relatively small aperture. There are about 270 living species in the family worldwide, with the highest diversity found in the tropical Indo-Pacific region. Some animals live in sand on coral reefs, others in the surf zone of beaches, and a few in muddy sand. All are carnivorous, feeding on marine worms. Augers, cones and turrids are collectively termed toxoglossa (poison tooth), based on their mode of feeding. They have a harpoon-like radula, which is used to inject venom and paralyze their prey. The triseriate auger shell is the most elongate of all coiled gastropods. An adult specimen, with all its whorls intact, is a rare find and highly desirable to collectors.

Distribution
Indo-West Pacific to Hawaii

Size
95 mm (3 ¾ in)

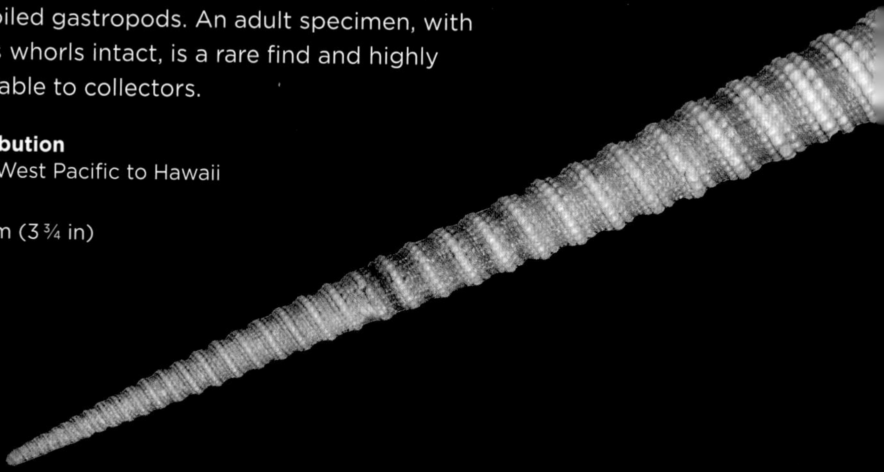

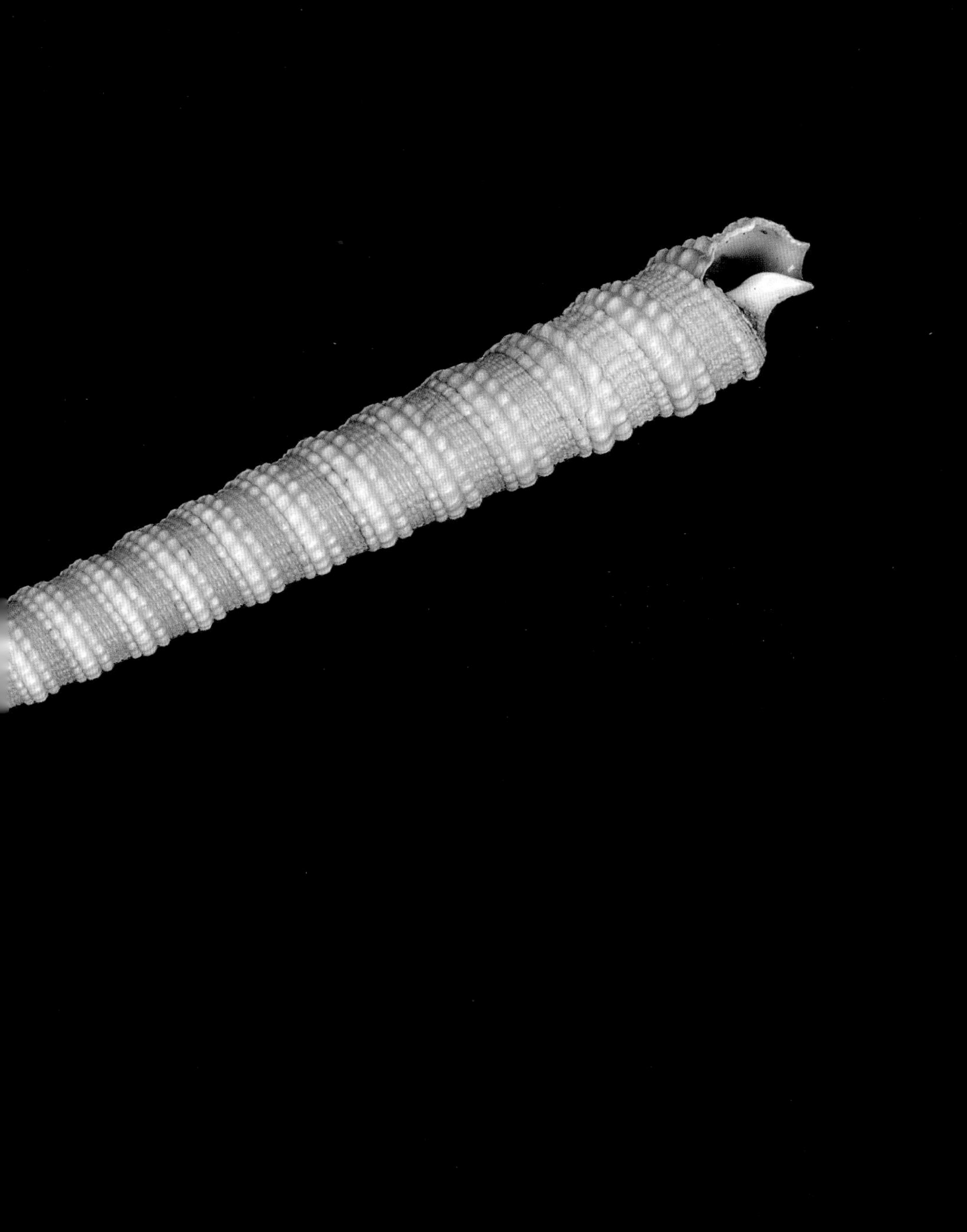

Ocean quahog

Arctica islandica

The longest-lived non-colonial animal known is a bivalve, an ocean quahog, that was dredged in 2006 off the coast of Iceland. The 507-year-old clam was initially named Ming after the Chinese dynasty in power when it was born. Later, it was renamed Hafrún (the mystery of the ocean) by the Icelandic researchers that discovered it. Its age was determined by counting growth lines in its shell, a method like counting tree rings. Molluscs increase the size of their shell by adding shell material at the edge of the mantle, and this produces the growth rings. The shell can also be used to look at what the marine environment was like in the past. Isotopes of oxygen and carbon in the shell can give us information about the sea temperature and salinity at the time each ring was formed. And the concentration of other elements in the shell, for example calcium or zinc, can be measured and compared to investigate marine pollution and ocean acidification changes through time.

Distribution
North Atlantic

Size
85 mm (3 ¼ in)

Black-lipped pearl oyster

Pinctada margaritifera

A pearl is formed when a foreign object such as a tiny piece of rock or shell, a parasite or even a fish, gets trapped inside the living mollusc. This becomes an irritant, and the animal reacts to it by covering the object with layers of the same material as the shell, thus creating a pearl. All species of molluscs are capable of producing pearls, however, only the ones that have a nacre or a mother-of-pearl shell interior, like the marine oysters or freshwater mussels, can create beautiful sought-after nacre pearls.

Distribution
Indian Ocean, western to central Pacific

Size
90 mm (3½ in)

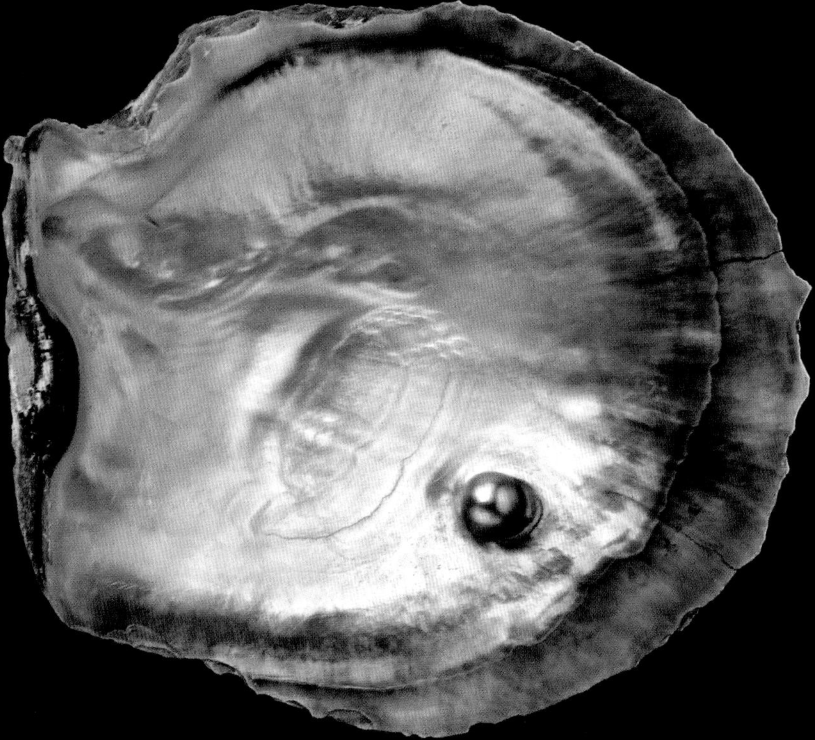

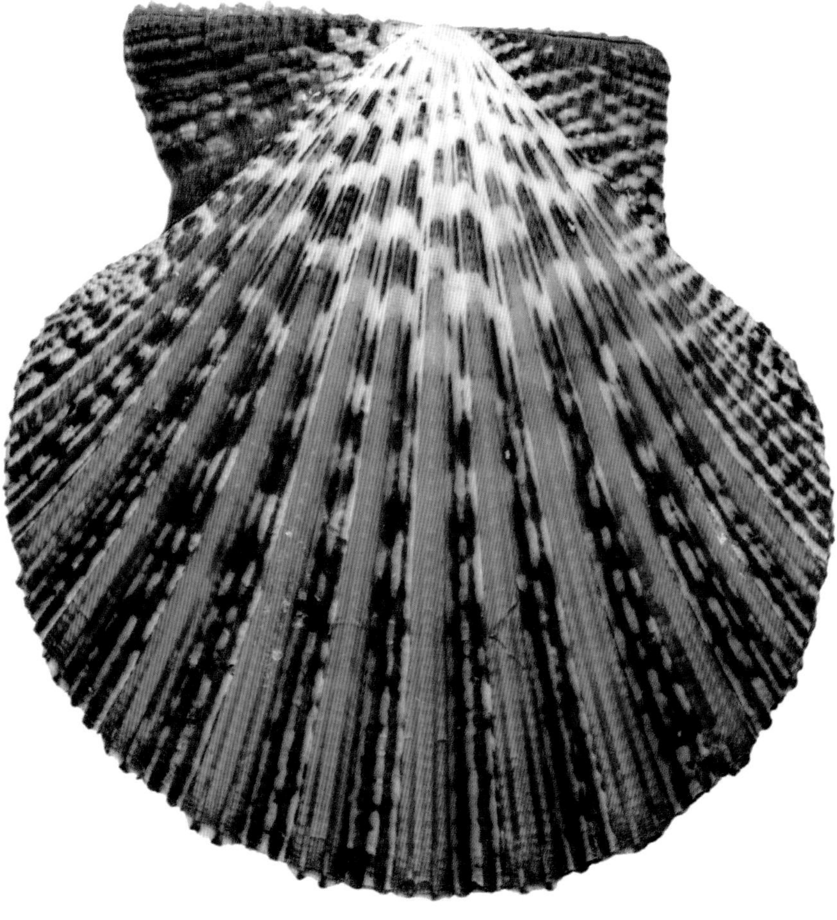

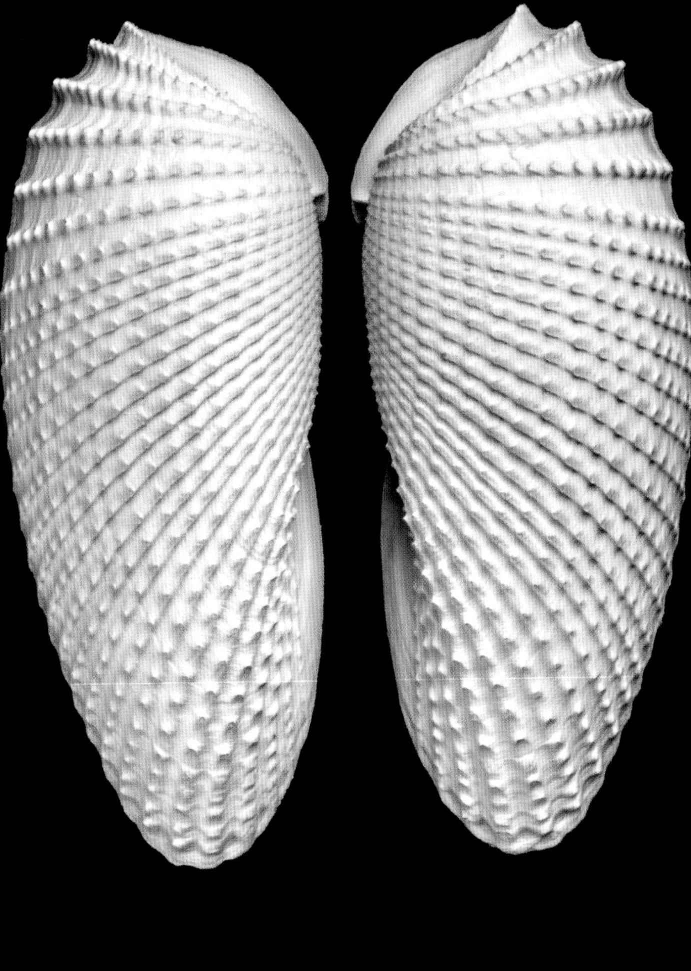

Northern quahog

Mercenaria mercenaria

Among the better-known edible bivalves is the northern quahog – it is the main ingredient of the famous clam chowder, a thick soup familiar to most Americans. It is an important species for commercial and recreational fisheries, as well as for aquaculture along the Atlantic coast of the USA. Native Americans have eaten this clam for thousands of years and the common name quahog is a modification of *poquaûhock* from the Narragansett people of Rhode Island. They also used the shell to produce beads that were cut or ground from the shell, and then pierced and threaded onto strings. These are called wampum and were used as money. Purple beads made from near the lip of the shell were particularly valuable.

Distribution
Eastern Canada to Gulf of Mexico

Size
90 mm (3 ½ in)

Saint James shell or king scallop

Pecten maximus

Scallop shells have long been used as cultural symbols, especially in religious contexts. It is the emblem of Saint James and of the pilgrims on the Way of Saint James (Camino de Santiago), a network of Catholic pilgrimages leading to the cathedral of Santiago de Compostela in Galicia, Spain. Saint James was one of the Twelve Apostles of Jesus and, according to tradition, his remains are buried in this cathedral. People from all over the world have been following these routes since the 9th century. During Medieval times, pilgrims would collect a scallop shell while at Santiago de Compostela as evidence of having made the journey; nowadays it is the symbol of the pilgrimage in general.

Distribution
European Atlantic Ocean, North Sea, and Mediterranean Sea

Size
80 mm (3 in)

Atlantic thorny oyster

Spondylus americanus

The spondylus, also known as thorny oysters, are filter-feeder bivalves found in the warm waters of most of the world's oceans. They cement themselves permanently to rocks or piers with the right valve. This means individuals can look quite different from others of the same species, because the shape of the shell adapts itself to the available space. It is common for them to be covered with algae, sponges or corals that settle and grow among the spines, creating a microhabitat and an ideal form of camouflage. They are found in a variety of colours and their sharp spines protect them from predators.

Distribution
North Carolina (USA) to Brazil

Size
150 mm (6 in)

Bear paw clam, horse's hoof or strawberry clam

Hippopus hippopus

The bear paw clam is a species of giant clam that lives in coral sands near reefs to a depth of up to 6 m (20 ft), in the tropical Indo-Pacific. The shell is thick with prominent ribs that are white with reddish striped patterns. The juveniles are attached to the bottom of the sea by the byssus, a kind of silken anchor, which they lose as the animal grows. The adults, unlike other species of giant clams, are free-living. Although this species is protected under national and international laws, it is still collected for food and shell craft in several Pacific and Asian countries. For this reason, populations continue to decline and local extinctions have even been recorded.

Distribution
Tropical Indo-West Pacific

Size
135 mm (5 ¼ in)

Atlantic tiger lucina

Codakia orbicularis

Lucinids are a worldwide, largely tropical, group of marine clams, living in the intertidal zone down to depths of around 2,500 m (8,200 ft). They are chemosymbiotic bivalves, possessing symbiotic bacteria in their gills from which they obtain much of their nutrition. The bacteria convert inorganic molecules, such as hydrogen sulphide, into organic compounds, providing the host with the necessary nutrients. This partnership between bivalve molluscs and sulphide- and methane-oxidizing bacteria allows the animals to exploit sulphide- and methane-rich habitats, such as deep-water hydrothermal vents and hydrocarbon seeps. They also mine the anoxic zones such as sands and muds, mangrove sediments, or even offshore sewage sites for sulphides. Atlantic tiger lucina is one of around 450 species of lucinids and lives burrowed in seagrass-covered sand from Florida to Brazil. It has a white, round, thick shell, sculptured by numerous curved fine ribs.

Distribution
Southeastern United States to Brazil

Size
65 mm (2 ½ in)

Lion's paw scallop

Nodipecten nodosus

Molluscs' eyes vary from simple pigment-cup eyes, such as those found in some limpets, to the camera-type eyes found in cephalopods, which in some species are capable of seeing images almost the same as a human eye would see. Scallops have the most acute vision of any bivalve. They have up to 200 small but well-developed eyes along the mantle edge between their two valves. Each of these eyes contains a tiny concave mirror to focus light, which is different from how most animals, including humans, see. Each eye can detect small changes in light intensity and movement, potentially alerting the scallop to approaching predators such as starfish, octopuses and sea snails. They can also see when the water is clouded with the microscopic algae and other particles they feed on. Scallops' eyes can be brightly coloured, like those of the lion's paw scallop which are a beautiful iridescent blue.

Distribution
Caribbean to Brazil, and tropical regions of western Africa

Size
80 mm (3 in)

Giant clam

Tridacna gigas

The giant clam is the largest species of bivalve mollusc and has the heaviest shell of all living molluscs, reaching 1.2 m (4 ft) and weighing 300 kg (660 lb). It feeds on plankton but also gets food from zooxanthellae – symbiotic algae that live inside the clam's tissues. The algae absorb the energy from the sun and turn it into food that it shares with the clam. It is the sugars and proteins produced by the algae that enable the clam to reach its enormous size. In return, giant clams provide the algae with a safe home. The bright colour patterns in the mantle of the clam, unique to each clam, are created by the zooxanthellae. Giant clams are rare in many areas because they have been overfished, but they are now being grown commercially for food to reduce harvesting of wild populations.

Distribution
Indo-Pacific

Size
310 mm (12 ¼ in)

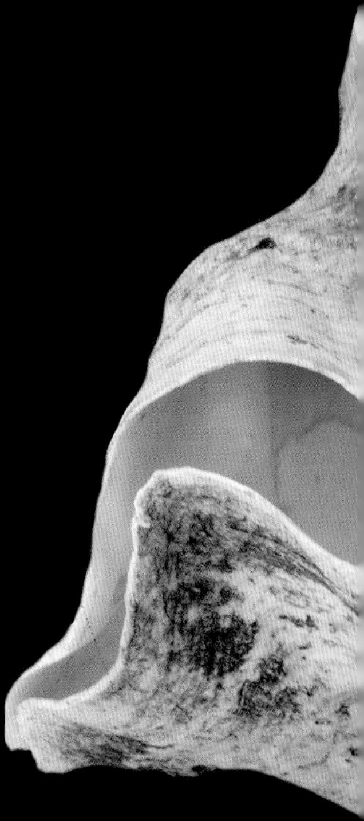

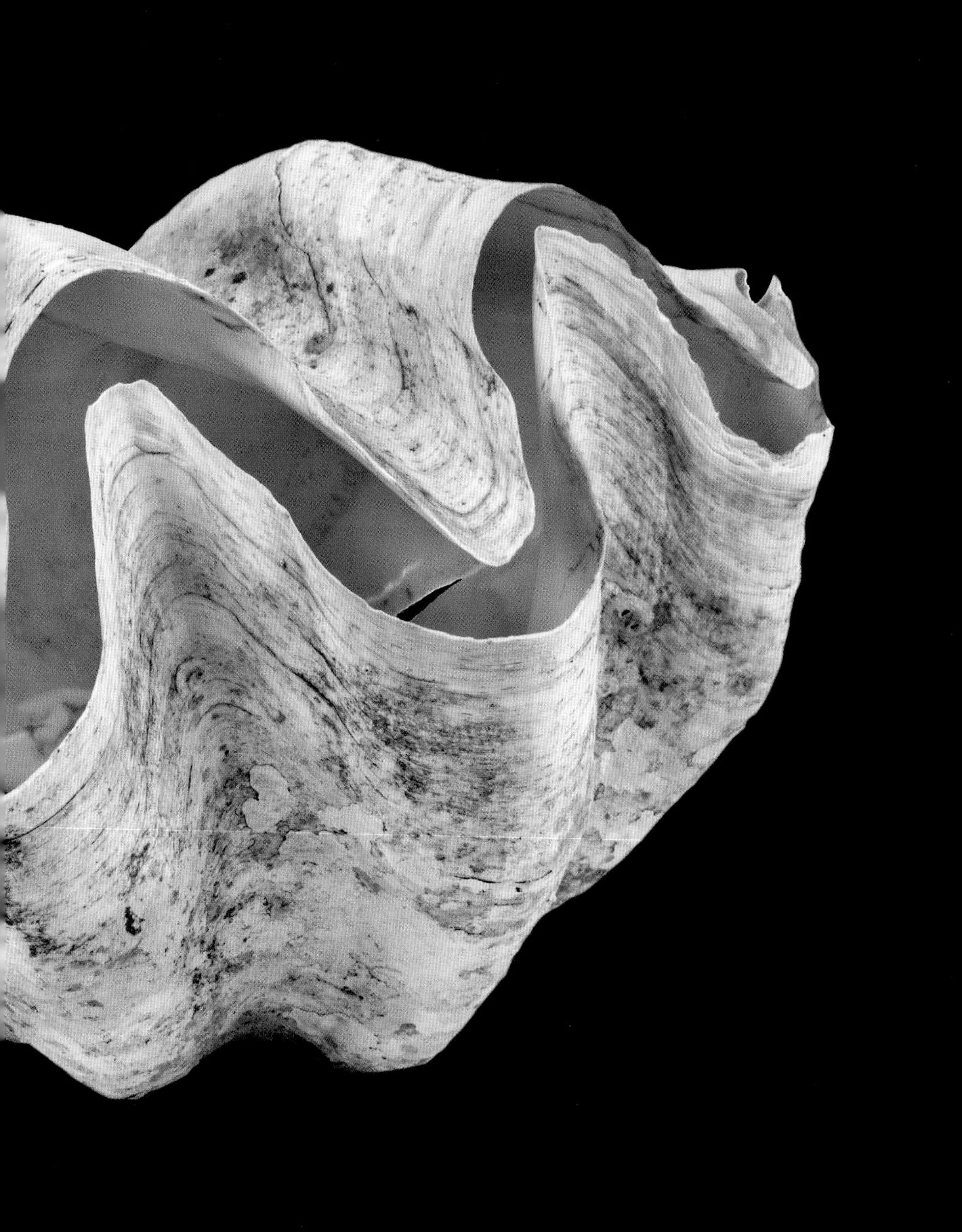

Frond oyster

Dendostrea frons

The frond oyster is a small marine bivalve that lives attached to gorgonians or soft corals, rocks and other hard substrates. Its shell varies in shape, depending on the appearance of the substrate it is attached to. When growing on gorgonians, it becomes elongated in form, and the left valve develops claw-like projections, called claspers, that hold on to the coral. When growing on rocks, the shape of the shell is more ovate. Frond oysters belong to the same family as edible oysters, the Ostreidae.

Distribution
North Carolina, USA to Brazil

Size
60 mm (2 ½ in)

Cockscomb pearl mussel

Cristaria plicata

The very first evidence of successful pearl production was in the 5th century BC by the Chinese. Small objects of ivory, ceramic, shell, lead, or clay, in the shape of spheres, hearts, fish or, most popular, small Buddhas, were inserted between the body and shell of the freshwater pearl mussel. The mussels were then returned to the water and left for about a year, after which the shells were reopened to reveal the objects coated in thin layers of mother-of-pearl. The small-pearlized items were then cut out of the shell and sold in shops. The tradition carries on today and this species of freshwater mussel is still one of the most important for pearl production in China.

Distribution
China, Japan and Vietnam

Size
70 mm (2 ¾ in)

Ox heart clam or heart cockle

Glossus humanus

The ox heart clam is a European marine bivalve. It is found from depths of 7 to 250 m (23 to 820 ft) off the Atlantic coastlines of Europe and North Africa half buried in soft, muddy substrates. It is a suspension feeder, feeding on plankton and other microscopic particles that it filters out with its gills. The shape of its large shell is very unusual – each valve has a coil resembling a gastropod shell. It is popularly known as the ox heart or heart cockle because of its characteristic heart shape when viewed from the side. The outer layer of the shell, called the periostracum, is usually dark brown or olive green, and frequently covered with short hair-like structures; the inner layer is beige or light orange, with fine radial lines.

Distribution
Norway to Mediterranean

Size
65 mm (2 ½ in)

Queen scallop

Aequipecten opercularis

The scallop is one of the few bivalve molluscs that can swim, and when threatened by a predator can move very quickly. It swims by filling its mantle cavity with water, while keeping the valves as wide apart as possible. Then it quickly claps the valves shut by contracting its large adductor muscle, forcing the water out in a jet. The opening and rapid closing of its valves propels the animal through the water in short bursts. It can control its direction of movement by directing the water jet with the mantle as the water is forced out.

Distribution
East Atlantic Ocean and Mediterranean

Size
55 mm (2 ¼ in)

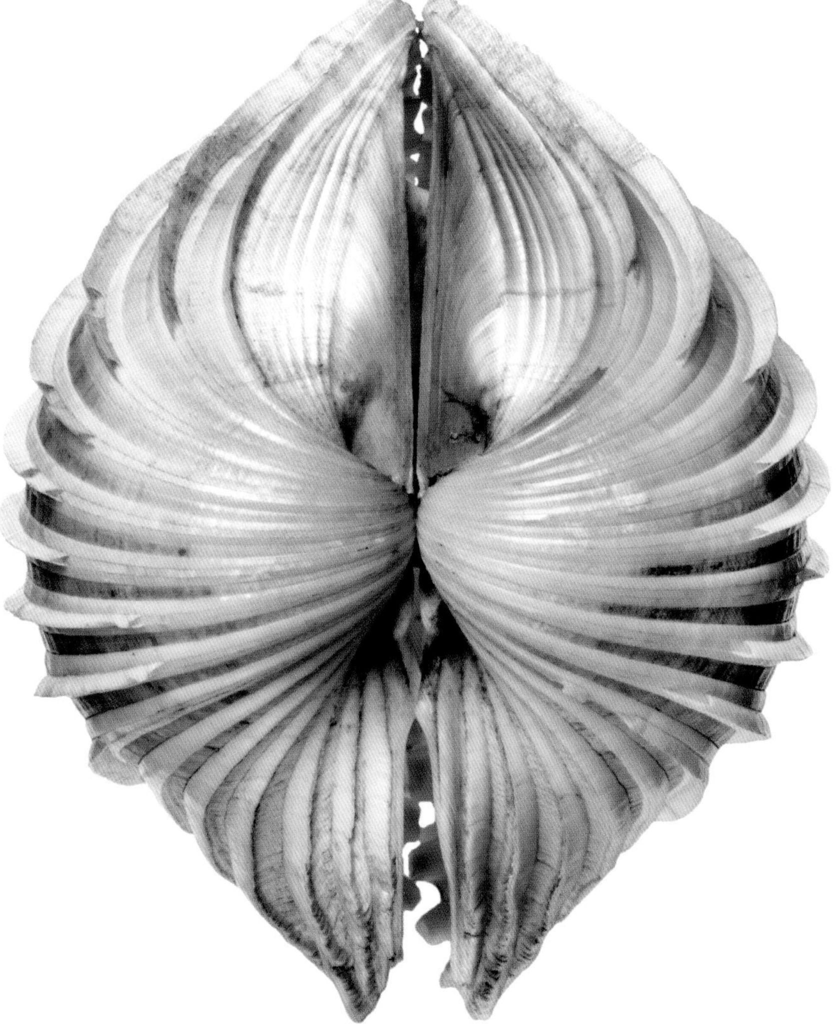

Great ribbed cockle

Cardium costatum

The great ribbed cockle is a beautiful bivalve with a striking shell that is easily recognisable because of the raised ribs on each valve. The shells are thin, inflated and white, with some orange-brown coloration between the ribs. It is one of the most emblematic species of the West African coast, where it lives buried in soft substrates in semi-sheltered bays as well as open coastlines and, occasionally, it washes ashore by the thousands after a storm. The cockles are a large group of species found worldwide in shallow sand and mud habitats where they burrow. They are able to jump by pressing their large and powerful foot against the substrate and straightening it, thus causing them to leap off the bottom.

Distribution
Senegal to Angola

Size
55 mm (2 ¼ in)

Wedding cake venus

Bassina disjecta

The wedding cake venus has a beautiful and distinctive shell with characteristic concentric ribs. It is usually white with pink on the lower side of the ribs. Like most bivalves it is a filter feeder, feeding on tiny particles of organic matter suspended in the surrounding water. The bivalve draws in water through a pair of short tubes, the siphons, and the microscopic food is filtered by the mucus-covered gills where it becomes trapped. The plankton is then removed from the gill to the mouth by the wave-like action of countless hair-like cilia.

Distribution
New South Wales to South Australia

Size
60 mm (2 ½ in)

Windowpane oyster

Placuna placenta

The windowpane oyster is one of the most flattened bivalves. Its greatly compressed valves hardly leave room for the animal, which lies on the right valve. It is abundant in quiet lagoons, bays and mangroves, on muddy bottoms in the tropical Indo-West Pacific. It feeds mostly on plankton and organic detritus. The shell is thin, brittle and translucent, with a pearlescent appearance and is nearly circular in outline. Their common name is derived from their long-standing use in windowpanes, in place of glass, in various Asian countries. They can still be seen in the windows of old houses in the Philippines, where they are fished in large numbers and manufactured into screens, lamp shades and ornaments.

Distribution
Tropical Indo-West Pacific

Size
150 mm (6 in)

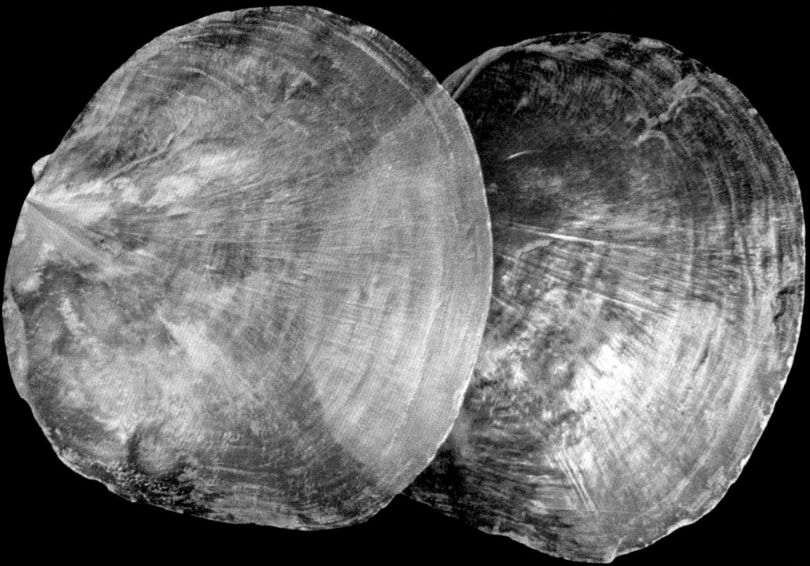

Common watering pot

Verpa penis

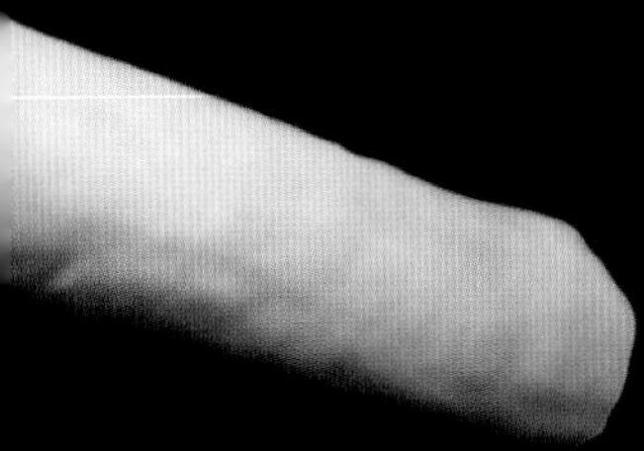

The watering pot is a very distinctive and unusual mollusc. In the early stages of development, the animal has a two-valve shell like other bivalves. As the animal starts to grow, it secretes a long calcareous tube that covers the body, in which its vestigial shell becomes embedded. At the front end is a curious structure that looks like the sprinkler of a watering can, hence its common name. This part is kept buried in soft, muddy sand, while the narrow open end of the tube is kept at the surface level of the sediment. The watering pot is a filter feeder and is found in the shallow waters of the tropical Indo-Pacific region. There are 15 known living species in this family worldwide.

Distribution
Indo-West Pacific

Size
80 mm (3 in)

Cock’s comb mussel

Lopha cristagalli

The shells of some bivalves are thin and fragile, but others, such as this, are thick and strong. The valves are hinged together by their upper edges and the animal must be able to close them quickly and securely to protect themselves from predators. For this reason, the two halves of a bivalve shell match perfectly in most cases and, when shut, they are impenetrable. The cock’s comb mussel has a very distinct shell, with strong folds at acute angles and a characteristic zigzag pattern that resembles a rooster’s comb. These molluscs are epifaunal suspension feeders – they live in shallow subtidal waters, attached with clasping spines to rocks, corals or other substrates.

Distribution
Indo-Pacific

Size
80 mm (3 in)

Black hammer oyster

Malleus malleus

One species that had an extraordinary appeal for 18th century collectors was the hammer oyster, and it is not unusual to see this bivalve in portraits on a table of shells with other curiosities. The attraction was almost certainly due to the distinctive and irregularly T-shaped shell, which is suggestive of a hammer. The black hammer oyster is found in shallow waters near coral reefs, vertically embedded and partially buried in coarse sand or sandy gravel, where it is anchored to the substrate by its byssus threads. The outer side of the shell is often thickly covered with seaweed, algae and other organisms. The interior is black with a patch of dark, iridescent nacre at the spot where the animal lives.

Distribution
Indo-Pacific

Size
190 mm (7 ½ in)

Spiny venus

Hysteroconcha lupanaria

The spiny venus is one of the most beautifully ornamented clams. The shell is characterized by its violet and purple-white coloration together with two rows of long, straight or slightly curved spines. It is very uncommon for a bivalve, apart from the thorny oysters, to have long spines; in fact, these are the longest spines of any bivalve. The spines are directed towards the surface, surrounding the animal's siphons, presumably protecting them from predators. They are filter feeders, living in sandy substrates in shallow waters.

Distribution
Baja California to Peru

Size
75 mm (3 in) with spines

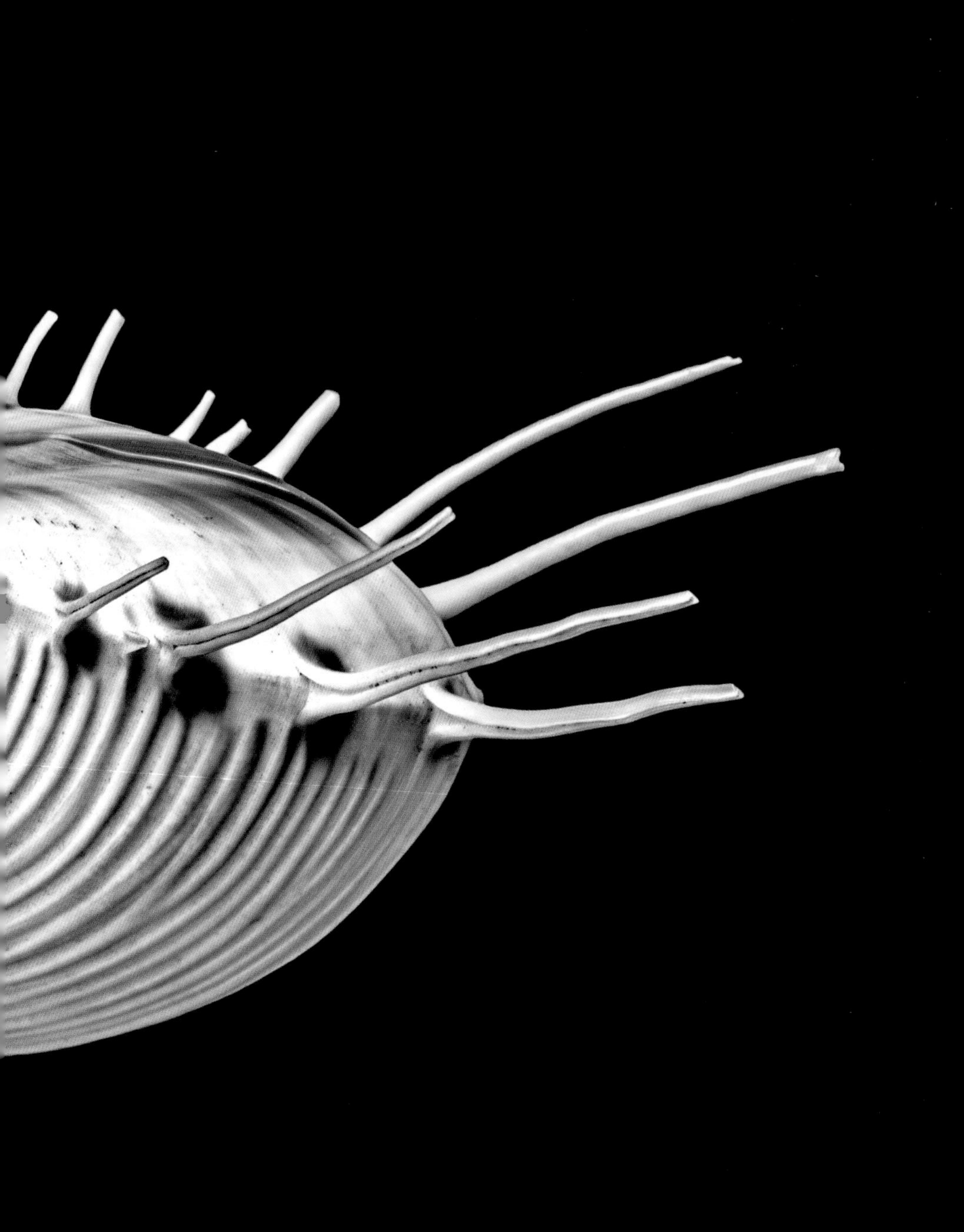

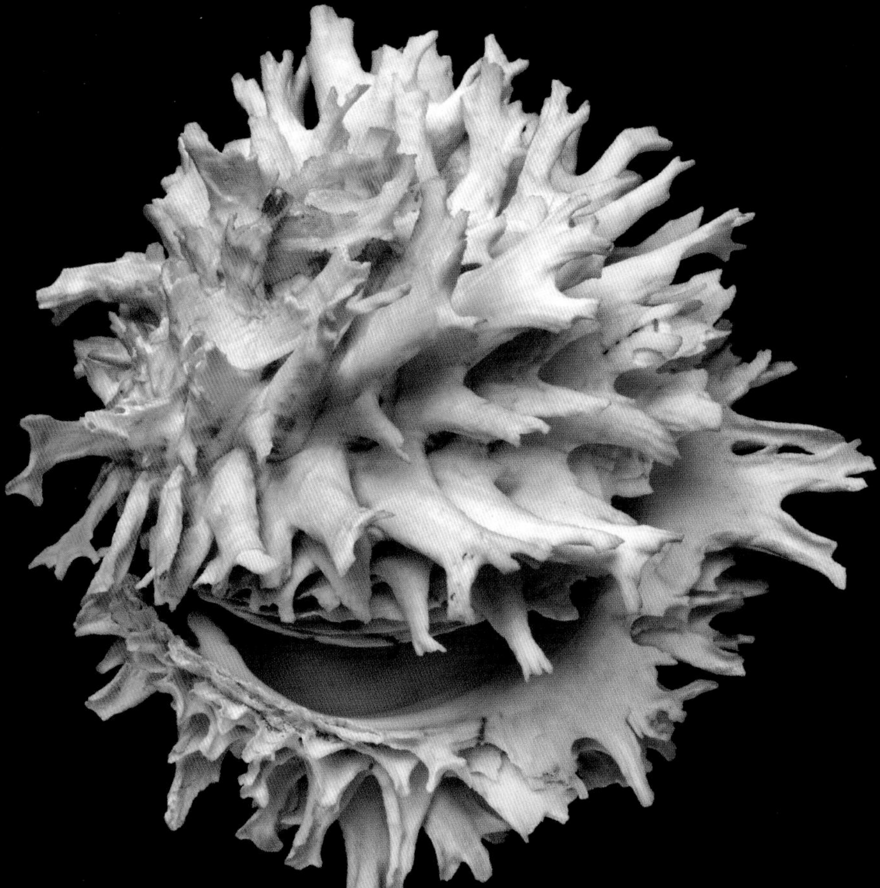

Lazarus jewel box

Chama lazarus

Jewel box shells are a relatively small species group found in temperate and tropical waters worldwide. They get their name from the upper valve of the shell, which forms a lid over the larger lower valve. These saltwater clams are reef-dwellers, living in shallow, clean warm waters, cementing themselves to a hard substrate such as corals, rocks or other shells. The Lazarus jewel box has elongated foliated spines, which are probably the largest in this family. The spines increase the effective volume of the shell and provide holdfasts for encrusting organisms. The shell can be variable in form and colour – white, yellow, red, pink or even a combination of colours.

Distribution
Red Sea to Indo-West Pacific

Size
80 mm (3 in) with spines

Zigzag venus

Lioconcha castrensis

The zigzag venus has an unmistakable shell, with large brown or black mountain peak markings on its ivory surface. It is common on shallow, sandy bottoms in coral reef areas throughout the Indo-Pacific. The venus clams are one of the largest groups of bivalves, with a global distribution, and come in an incredible variety of shapes and sizes. The animals burrow into the sediment, leaving their siphons, and often the upper surface of the shell, exposed. They are suspension-feeders, filtering algae and organic sediment from the water.

Distribution
Red Sea to Indo-Pacific

Size
50 mm (2 in)

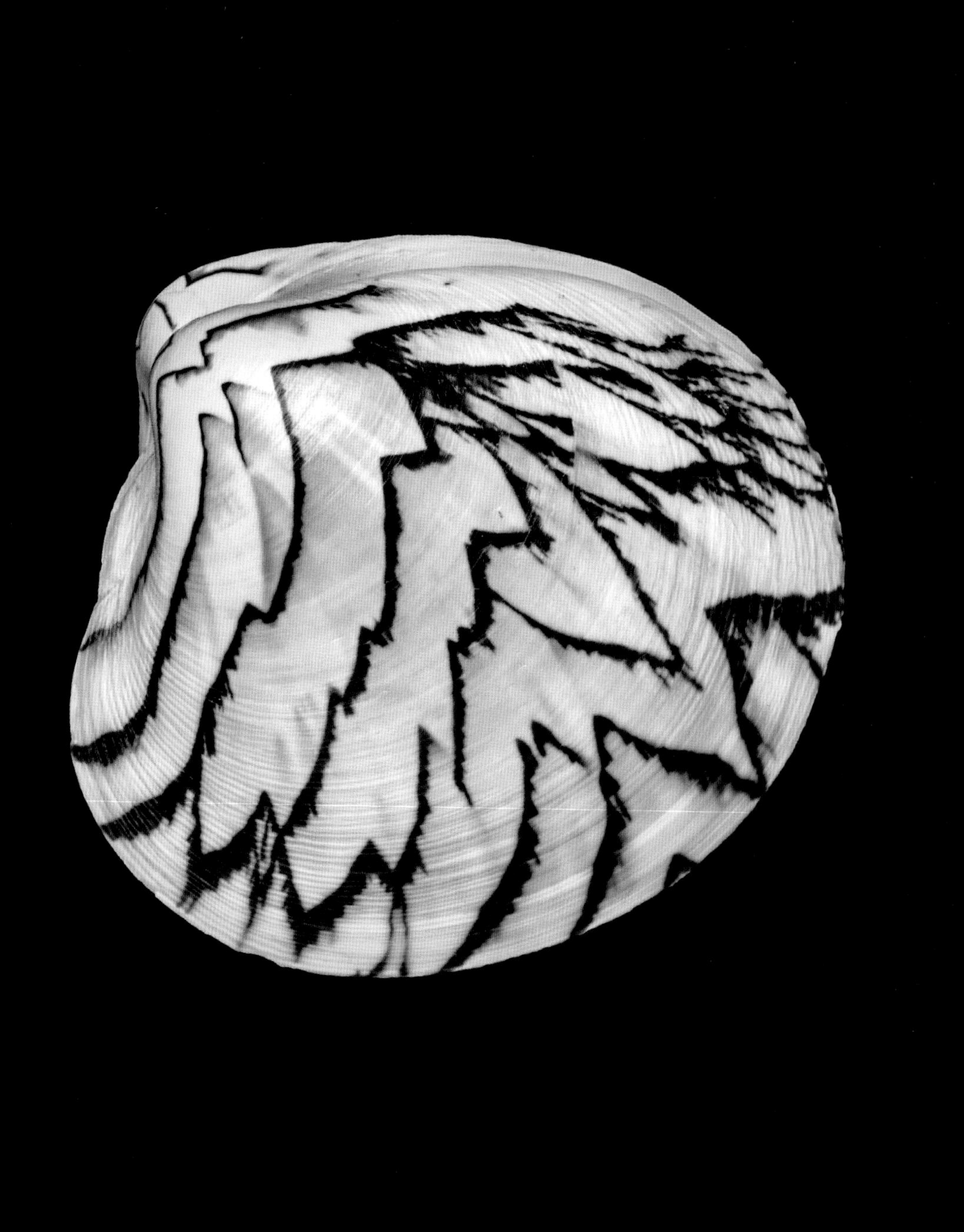

Heart cockle

Corculum cardissa

The heart-shaped shell of *Corculum cardissa* has one side concave and the other convex. It is a common species that lives near coral reefs and rests horizontally with its flatter side down on sandy bottoms in shallow water. Like the related giant clams (genus *Tridacna*), the heart cockles have a symbiotic relationship with algae that live within its tissues. However, unlike the giant clam, it does not expose the tissues directly to the sunlight. Instead, the thin shell of the heart cockle has tiny translucent windows that allow light to reach inside the shell. In this way, the algae can photosynthesize and provide nutrients to the animal, and in return the algae have a safe place to live.

Distribution
Red Sea to Indo-Pacific

Size
40 mm (1½ in)

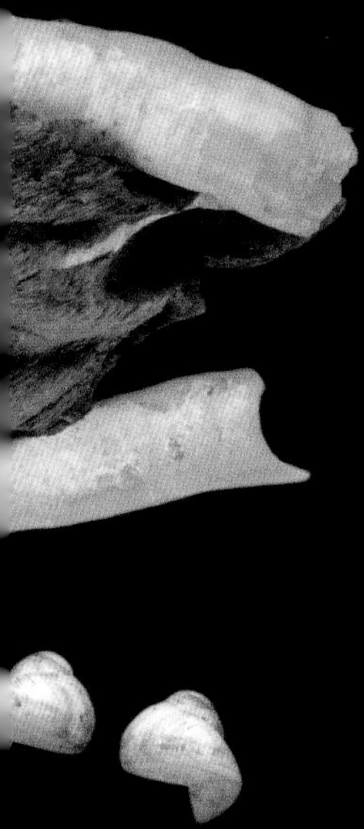

Naval shipworm

Teredo navalis

Shipworms are bivalves with a long, worm-like body that can reach up to 1 m (3 ft) in length. In contrast, the shell is very small, helmet-shaped and covers only the anterior end of the animal. These molluscs are among the most destructive invertebrates in the world, causing a great deal of damage to wooden ships and waterfront structures. They bore deep tunnels into wood with the sharp edge of their shell, after which the animal secretes a chalky tube that lines its burrow protecting the body. Their wood-eating habits have been known since Greek and Roman times and have even earnt shipworms a prominent place in maritime history for 'sinking more ships than pirates'. In 1502, Christopher Columbus's fourth voyage to the Americas came to a disastrous end when several of his ships sank due to damage caused by shipworms. The same happened to Francis Drake's flagship, the *Golden Hind,* which foundered at her anchorage in the River Thames at Woolwich, London.

Distribution
Cosmopolitan

Size
8 mm (⅓ in)

Chambered nautilus

Nautilus pompilius

The nautiloids are the only group of living cephalopods with true external shells. The nautilus has a shell made up of chambers, which are connected by a tube called the siphuncle. The animal, with around 90 tentacles, occupies only the last chamber of the shell, into which it can withdraw completely. As it grows, it creates a new larger chamber, moving its body into the new space and sealing off the vacated one. Nautilus shells had been carved, etched and engraved for many centuries, but rose to prominence in the 17th century when craftsmanship reached new heights of design in Europe, particularly in the Netherlands. The carved shells, mounted as cups in silver or gold, were popular items in cabinets of curiosities and among the most coveted natural history objects.

Distribution
Indo-Pacific

Size
160 mm (6 ¼ in)

Paper nautilus

Argonauta argo

Despite appearances, the argonauts, or paper nautilus, are more closely related to the common octopus than to the chambered nautilus. Their shell is, in fact, not a shell at all, but a protective egg case. The female argonaut has two modified arms that secrete the paper-thin white structure, building it in the shape of a little boat, where she lays and keeps her eggs as she swims. She carries her thousands of minute eggs around with her, inhabiting the cradle together with her brood, protecting them until they hatch. After hatching, the fragile shell-like case is discarded.

Distribution
Worldwide in warm waters

Size
230 mm (9 in)

Ram's horn squid or common spirula

Spirula spirula

This small species of deep-water squid-like cephalopod occurs worldwide in warm waters. The animal has rarely been seen alive, although the small internal shell often washes up on beaches and shores, and is extremely light, coiled and chambered. It has a daily vertical migration, rising from depths ranging from 500 to 1,000 m (1,640 to 3,280 ft) during the day to feed in shallower waters at night. This is only possible because they use their shell to control their buoyancy, as with the *Nautilus*.

Distribution
Worldwide in warm waters

Size
25 mm (1 in)

West Indian green chiton

Chiton tuberculatus

Chitons are a group of marine molluscs distinguished by having a series of eight articulated and overlapping valves attached to the back of the soft-bodied animal. These valves are surrounded and held together by a tough muscular band known as a girdle, which may be smooth or ornamented with spines, bristles or knobs. The overlapping valves allow chitons to mould to any irregularities in the rock surface. When detached the chitons curl up, like a woodlouse or an armadillo, to protect their soft body. The common West Indian green chiton is one of the largest chitons occurring in the Caribbean, reaching up to 100 mm (4 in) long, where it lives on rocky shores.

Distribution
Florida to Venezuela, West Indies

Size
70 mm (2 ¾ in)

Elephant tusk

Dentalium elephantinum

The scaphopods or tusk shells constitute a small class of molluscs that have curved shells similar in shape to miniature elephant tusks. All species are marine and can be found in both deep and shallow waters. They live buried in sand or mud with just the tip of the narrow end of the shell sticking out of the sea floor. They have no head, no eyes, no gills, not even a heart, and feed on microscopic organisms that they catch with their sticky tentacles, then drag into their mouth. Continuous contractions of the foot, which emerges from the wide end of the shell, help to circulate both water and blood through the body. Oxygen is extracted from the water through cells situated in the mantle. The bright green elephant tusk is a relatively common shell that can grow to 100 mm (4 in) in length.

Distribution
Red Sea to Philippines

Size
90 mm (3 ½ in)

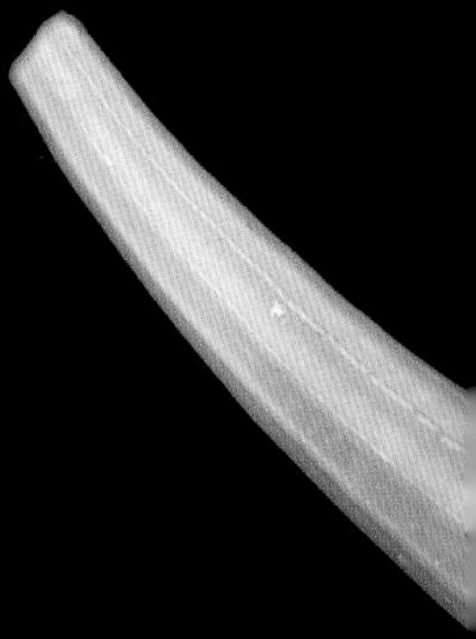

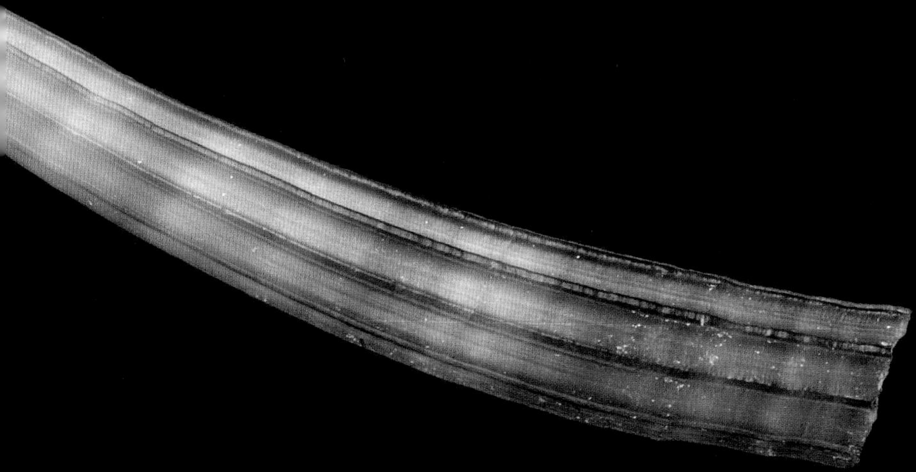

Glossary

Abyssal: occurring at ocean depths ranging from 4,000 to 6,500 m (13,123 to 21,324 ft).

Conchiolin: a horny organic component of the shell.

Epifaunal: living on the surface of a substrate in an aquatic environment.

Malacology: the study of molluscs.

Mantle: a thin layer of tissue that covers the soft part of the body. It secretes the shell and periostracum.

Mantle cavity: a space between the mantle and the body that contains the respiratory organs.

Mollusc: a soft-bodied animal without an internal skeleton, usually covered by a shell.

Pelagic: living in open ocean waters.

Radula: a flexible ribbon to which rows of rasping teeth are attached. A feeding structure unique to molluscs but absent in bivalves.

Siphon: a tubular organ or an extension of the mantle through which water passes.

Substrate: the bottom or supporting surface, such as the sea floor, rocks and coral.

Symbiotic: different species living together in an association that harms neither participant.

Toxoglossa: radular teeth modified for injecting venom into prey.

Umbilicus: the central cavity at the base of the coiled shell, around which the whorls coil.

Valve: one of the halves of a bivalve shell.

Varix (plural varices): prominent raised ridges on the surface of the shell – formed originally at the aperture during rest periods in the growth of the shell.

Whorl: one complete turn of a gastropod shell.

Zooxanthellae: microscopic algae that live inside the clam's tissues.

Index